RealTime Physics
Active Learning Laboratories

Module 3

Electric Circuits

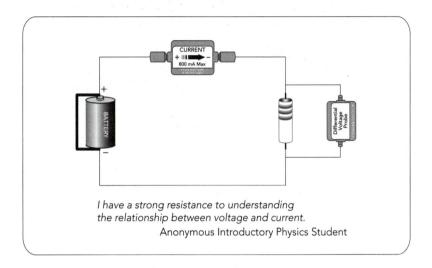

*I have a strong resistance to understanding
the relationship between voltage and current.*
Anonymous Introductory Physics Student

David R. Sokoloff
Department of Physics
University of Oregon

Priscilla W. Laws
Department of Physics
Dickinson College

Ronald K. Thornton
Center for Science and Math Teaching
Departments of Physics and Education
Tufts University

with contributing authors:

Jeffrey Marx
McDaniel College

Molly Johnson
Greeley, Colorado

John Wiley & Sons, Inc.

ACQUISITIONS EDITOR	Stuart Johnson
PUBLISHER	Kaye Pace
PROJECT EDITOR	Geraldine Osnato
MARKETING MANAGER	Robert Smith
PRODUCTION DIRECTOR	Pam Kennedy
PRODUCTION EDITOR	Sarah Wolfman-Robichaud
SENIOR DESIGNER	Kevin Murphy
ILLLUSTRATION EDITOR	Anna Melhorn
COVER PHOTO	James Fraher/Image Bank/Getty Images

This book was set in Palatino by GTS Companies/York, PA Campus and printed and bound by Courier/Westford. The cover was printed by Phoenix Color.

This book is printed on acid free paper. ∞

Printed in the United States of America

10 9 8 7 6 5 4

ACKNOWLEDGMENTS

RealTime Physics Module 3: Electric Circuits could not have been developed without the hardware and software development work of Stephen Beardslee, Lars Travers, Ronald Budworth, and David Vernier. We are indebted to numerous college, university, and high school physics teachers, and especially Curtis Hieggelke (Joliet Junior College), John Garrett (Sheldon High School), and Maxine Willis (Gettysburg High School) for beta testing earlier versions of the laboratories with their students. We thank Mary Fehrs, Matthew Moelter, Gene Mosca, and Sharon Schmalz for beta testing the labs and for their invaluable suggestions and corrections in the final stages of editing.

At the University of Oregon, we especially thank Dean Livelybrooks for supervising the introductory physics laboratory, for providing invaluable feedback, and for writing some of the homework solutions for the *Teachers' Guide*. Frank Womack, Dan DePonte, Sasha Tavenner, and all of the introductory physics laboratory teaching assistants provided valuable assistance and input. We also thank the faculty at the University of Oregon (especially Stan Micklavzina), Tufts University, and Dickinson College for their input, and for assisting with our conceptual learning assessments. Finally, we could not have even started this project if not for our students' active participation in these endeavors.

This work was supported in part by the National Science Foundation under grant number DUE-9455561, *"Activity Based Physics: Curricula, Computer Tools, and Apparatus for Introductory Physics Courses,"* grant number USE-9150589, *"Student Oriented Science,"* grant number DUE-9451287, *"RealTime Physics II: Active University Laboratories Based on Workshop Physics and Tools for Scientific Thinking,"* grant number USE-9153725, *"The Workshop Physics Laboratory Featuring Tools for Scientific Thinking,"* and grant number TPE-8751481, *"Tools for Scientific Thinking: MBL for Teaching Science Teachers,"* and by the Fund for Improvement of Post-secondary Education (FIPSE) of the U.S. Department of Education under grant number G008642149, *"Tools for Scientific Thinking,"* and number P116B90692, *"Interactive Physics."*

This project was supported, in part, by the National Science Foundation. Opinions expressed are those of the authors and not necessarily those of the foundation.

PREFACE

Development of the series of *RealTime Physics* (*RTP*) laboratory guides began in 1992 as part of an ongoing effort to create high-quality curricular materials, computer tools, and apparatus for introductory physics teaching.[1] The *RTP* series is part of a suite of *Activity-based Physics* curricular materials that include the *Tools for Scientific Thinking* laboratory modules,[2] the *Workshop Physics* Activity Guide,[3,4] and the *Interactive Lecture Demonstration* series.[5] The development of all of these curricular materials has been guided by the outcomes of physics education research. This research has led us to believe that students can learn vital physics concepts and investigative skills more effectively through guided activities that are enhanced by the use of powerful microcomputer-based laboratory (MBL) tools.

In the past ten years new MBL tools—originally developed at Technical Education Research Centers (TERC) and at the Center for Science and Mathematics Teaching, Tufts University—have become increasingly popular for the real-time collection, display, and analysis of data in the introductory laboratory. MBL tools consist of electronic sensors, a microcomputer interface, and software for data collection and analysis. Sensors are now available to measure such quantities as force, sound, magnetic field, current, voltage, temperature, pressure, rotary motion, acceleration, humidity, light intensity, pH, and dissolved oxygen.

MBL tools provide a powerful way for students to learn physics concepts. For example, students who walk in front of an ultrasonic motion sensor while the software displays position, velocity, or acceleration in real time more easily discover and understand motion concepts. They can see a cooling curve displayed instantly when a temperature sensor is plunged into ice water, or they can sing into a microphone and see a pressure vs. time plot of sound intensity.

MBL data can also be analyzed quantitatively. Students can obtain basic statistics for all or a selected subset of the collected data, and then either fit or model the data with an analytic function. They can also integrate, differentiate, or display a fast Fourier transform of data. Software features enable students to generate and display *calculated quantities* from collected data in real time. For example, since mechanical energy depends on mass, position, and velocity, the time variation of potential and kinetic energy of an object can be displayed graphically in real time. The user just needs to enter the mass of the object and the appropriate energy equations ahead of time.

The use of MBL tools for both conceptual and quantitative activities, when coupled with recent developments in physics education research, has led us to expand our view of how the introductory physics laboratory can be redesigned to help students learn physics more effectively.

Common Elements in the *Realtime Physics* Series

Each laboratory guide includes activities for use in a series of related laboratory sessions that span an entire quarter or semester. Lab activities and homework assignments are integrated so that they depend on learning that has occurred during the previous lab session and also prepare students for activities in the next session. The major goals of the *RealTime Physics* project are (1) to help students acquire an understanding of a set of related physics concepts; (2) to provide students with direct experience of the physical world by using MBL tools for real-time data collection, display, and analysis, (3) to enhance traditional laboratory skills; and (4) to reinforce topics covered in lectures and readings using a combination of conceptual activities and quantitative experiments.

To achieve these goals we have used the following design principles for each module based on educational research:

- The materials for the weekly laboratory sessions are sequenced to provide students with a coherent observational basis for understanding a single topic area in one semester or quarter of laboratory sessions.

- The laboratory activities invite students to construct their own models of physical phenomena based on observations and experiments.

- The activities are designed to help students modify common preconceptions about physical phenomena that make it difficult for them to understand essential physics principles.

- The activities are designed to work best when performed in collaborative groups of 2 to 4 students.

- MBL tools are used by students to collect and graph data in real time so they can test their predictions immediately.

- A learning cycle is incorporated into each set of related activities that consists of prediction, observation, comparison, analysis, and quantitative experimentation.

- Opportunities are provided for class discussion of student ideas and findings.

- Each laboratory comes with a pre-lab warm-up assignment, and with a post-lab homework assignment that reinforces critical physics concepts and investigative skills.

The core activities for each laboratory session are designed to be completed in two hours. Extensions have been developed to provide more in-depth coverage when longer lab periods are available. The materials in each laboratory guide are comprehensive enough that students can use them effectively even in settings where instructors and teaching assistants have minimal experience with the curricular materials.

The curriculum has been designed for distribution in electronic format. This allows instructors to make local modifications and reprint those portions of the materials that are suitable for their programs. The *Activity-Based Physics* curricular materials can be combined in various ways to meet the needs of students and instructors in different learning environments. The *RealTime Physics* laboratory guides are designed as the basis for a complete introductory physics laboratory program at colleges and universities, but they can also be used as the central component of a high school physics course. In a setting where formal lectures are given we recommend that the *RTP* laboratories be used in conjunction with *Interactive Lecture Demonstrations*.

The Electric Circuits Laboratory Guide

The primary goal of this *RealTime Physics Electric Circuits* guide is to provide students with a solid understanding of basic DC circuit concepts, and an introduction to AC circuits. A number of physics education researchers have documented that most students begin their studies with conceptions about the nature of circuits and circuit quantities like current and potential difference that can seriously inhibit their learning.[6,7]

McDermott and Shaffer[7] have documented the following student difficulties with circuits, among others: (1) failure to distinguish among concepts of current, potential difference, energy, and power; (2) lack of concrete experiences with real circuits; (3) failure to understand and apply the concept of a complete circuit; (4) belief that direction of current and order of elements matter; (5) belief that current is "used up" in a circuit; (6) belief that the battery is a constant current source; (7) failure to recognize that an ideal battery maintains a constant potential between its terminals; (8) failure to distinguish between branches connected in parallel across a battery and branches connected in parallel elsewhere; (9) failure to distinguish between potential and potential difference; and (10) difficulty in identifying series and parallel connections.

RealTime Physics Electric Circuits includes 8 labs:

Lab 1 (Batteries, Bulbs, and Current): In this lab students first explore the nature and definition of electric current. They then discover what is necessary for current to flow in a complete circuit, and practice connecting a variety of circuits. They use current probes to determine a model for the current flowing in a simple series circuit, and current and voltage probes to examine the potential difference maintained by a battery for various currents drawn from the battery.

Lab 2 (Current in Simple DC Circuits): Students explore the relationships between current in different parts of series and parallel circuits constructed first from bulbs and then from resistors. They use current and voltage probes to do their measurements.

Lab 3 (Voltage in Simple DC Circuits and Ohm's Law): Students explore the relationships between potential difference in different parts of series and parallel circuits constructed from bulbs and resistors. They also examine the internal resistance of a battery, and discover Ohm's law for a resistor by graphing the current through a resistor and the voltage across it simultaneously. They use current and voltage probes to do their measurements.

Lab 4 (Kirchhoff's Circuit Rules): First students examine how a multimeter is connected to measure current and voltage, and why it is connected that way. Then they learn how to measure resistance with a multimeter, and they use the multimeter to discover the rules for finding the equivalent resistance for series and parallel connections of resistors. Finally, they apply Kirchhoff's Circuit Rules to more complex circuits.

Lab 5 (Introduction to Capacitors and RC Circuits): Students first construct parallel-plate capacitors from aluminum foil sheets and examine the dependence of the capacitance on plate separation and plate area. Then they discover the rules for finding the equivalent capacitance for series and parallel connections of capacitors. Finally, they explore the transient behavior of RC circuits and the definition of time constant for an exponential decay.

Lab 6 (Introduction to Inductors and LR Circuits): Students first consider the behavior of a coil of wire (inductor) with a steady (DC) voltage across it. They then compare the behavior of a inductor to that of a resistor in a switching circuit in which the signal is turned on and off. They analyze the switching behavior of the inductor more quantitatively, then they examine the transient behavior of LR circuits and the time constant associated with these circuits.

Lab 7 (Introduction to Time-varying (AC) Signals): After first creating their own time-varying signals from a DC voltage, students explore the signals produced by a signal generator. Then they examine the behavior of a resistor with an AC signal applied to it. Finally, they explore in detail the behavior of capacitors and inductors with an applied AC signal, and compare this behavior to resistors.

Lab 8 (Introduction to AC Filters and Resonance): In this lab, students first examine the behavior of a simple capacitor circuit as a filter by applying different frequency signals. They then do the same for a simple inductor circuit. They explore the application of these observations to audio speaker design. Next they examine resonance in an RLC circuit and the application of such a circuit in a radio tuner. Finally they explore phase in RLC circuits.

On-Line Teachers' Guide

The *Teachers' Guide* for *RealTime Physics Electric Circuits* is available on-line at **http:/www.wiley.com/college/sokoloff.** This *Guide* focuses on pedagogical (teaching and learning) aspects of using the curriculum, as well as computer-based and other equipment. The *Guide* is offered as an aid to busy physics educators and does not pretend to delineate the "right" way to use the *RealTime Physics Electric Circuits* curriculum and certainly not the MBL tools. There are many right ways. The *Guide* does, however, explain the educational philosophy that influenced the design of the curriculum and tools and suggests effective teaching methods. Most of the suggestions have come from the college, university, and high school teachers who have participated in field testing of the curriculum.

The *On-line Teachers' Guide* has nine sections. Section I presents suggestions regarding computer hardware and software to aid in the implementation of this activity-based MBL curriculum. Sections II through IX present information about the eight laboratories. Included in each of these is information about the specific equipment and materials needed, tips on how to optimize student learning, answers to questions in the labs, and complete answers to the homework.

Experiment Configuration Files

Experiment configuration files are used to set up the appropriate software features to go with the activities in these labs. You will need the set of files that is designed for the software package you are using, or you will need to set up the files yourself. At this writing, experiment configuration files for *RealTime Physics Electric Circuits* are available for the Vernier Software and Technology *Logger Pro* (for Windows and Macintosh) and PASCO *Data Studio* (for Windows and Macintosh). Appendix A of this module outlines the features of the experiment configuration files for *RealTime Physics Electric Circuits*, as a guide to setting up configuration files for other software packages. For more information, consult the *On-line Teachers' Guide*.

Conclusions

RealTime Physics Electric Circuits has been used in a variety of different educational settings. Many university, college, and high school faculty who have used this curriculum have reported improvements in student understanding of circuit concepts. These comments are supported by our careful analysis of pre- and post-test data using the *Electric Circuit Conceptual Evaluation*, some of which have been reported in the literature.[8] Similar research on the effectiveness of *RealTime Physics*

Mechanics,[9] *Heat and Thermodynamics,* and *Light and Optics* also shows dramatic conceptual learning gains in these topic areas. We feel that by combining the outcomes of physics educational research with microcomputer-based tools, the laboratory can be a place where students acquire both a mastery of difficult physics concepts and vital laboratory skills.

References

1. Ronald K. Thornton and David R. Sokoloff, "RealTime Physics: Active Learning Laboratory," in *The Changing Role of the Physics Department in Modern Universities, Proceedings of the International Conference on Undergraduate Physics Education,* 1101–1118 (New York, American Institute of Physics, 1997).

2. Ronald K. Thornton and David R. Sokoloff, "Tools for Scientific Thinking—Heat and Temperature Curriculum and Teachers' Guide," (Portland, Vernier Software, 1993) and David R. Sokoloff and Ronald K. Thornton, "Tools for Scientific Thinking—Motion and Force Curriculum and Teachers' Guide," 2nd ed. (Portland, Vernier Software, 1992).

3. P. W. Laws, "Calculus-based Physics Without Lectures," *Phys. Today* **44** (12): 24–31 (1991).

4. Priscilla W. Laws, *Workshop Physics Activity Guide: The Core Volume with Module 1: Mechanics* (New York, Wiley, 1997).

5. David R. Sokoloff and Ronald K. Thornton, "Using Interactive Lecture Demonstrations to Create an Active Learning Environment," *The Physics Teacher* **27** (6):340 (1997).

6. L. C. McDermott, "Millikan Lecture 1990: What We Teach and What Is Learned—Closing the Gap," *Am. J. Phys* **59,** 301 (1991).

7. Lillian C. McDermott and Peter S. Shaffer, "Research as a Guide to Curriculul Development: An Example from Introductory Electricity. Part I: Investigation of Student Understanding," *Am. J. Phys.* **60,** 994 (1992).

8. David R. Sokoloff, "Teaching Electric Circuit Concepts Using Microcomputer-based Current and Voltage Probes," chapter in *Microcomputer-Based Labs: Educational Research and Standards,* Robert F. Tinker, ed., *Series F, Computer and Systems Sciences,* **156,** 129–146 (Berlin, Springer-Verlag, 1996).

9. Ronald K. Thornton and David R. Sokoloff, "Assessing Student Learning of Newton's Laws: The *Force and Motion Conceptual Evaluation* and the Evaluation of Active Learning Laboratory and Lecture Curricula," *Am. J. Phys.* **66:** 338–352 (1998).

REALTIME PHYSICS ELECTRIC CIRCUITS
Table of Contents

Name_____   Date_____

Pre-Lab Preparation Sheet for Lab 1—Batteries, Bulbs, and Current

(Due at the beginning of lab)

Directions:
Read over Lab 1 and then answer the following questions about the procedures.

1. Explain why in Activity 1-1 the angle irons will be charged in several different ways.

2. Sketch below one arrangement of the battery, bulb, and wire (other than the one in Figure 1-3), which you will try in Activity 1-2.

3. At this time, which model for current in Figure 1-6 do you favor? Why?

4. Describe briefly how you will test the models for current. What device will you use to measure current?

5. What is the symbol for a battery? A light bulb?

LAB 1:
BATTERIES, BULBS, AND CURRENT*

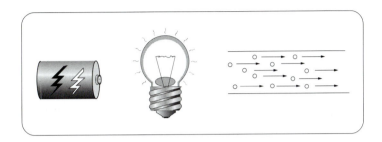

*You cannot teach a man anything; you
can only help him to find it within himself.*

—Galileo

OBJECTIVES

- To understand how a potential difference (voltage) can cause an electric current through a conductor.

- To learn to design and construct simple circuits using batteries, bulbs, wires, and switches.

- To learn to draw circuit diagrams using symbols.

- To understand the measurement of current and voltage using microcomputer-based probes.

- To understand currents at all points in simple circuits.

OVERVIEW

In the following labs, you are going to discover and extend theories about electric charge and potential difference (voltage), and apply them to electric circuits. What you learn will be one of the most practical parts of the whole physics course, since electric circuits form the backbone of twentieth-century technology. Without an understanding of electric circuits we wouldn't have lights, air conditioners, automobiles, telephones, TV sets, dishwashers, computers, or photocopying machines.

A *battery* is a device that generates an electric potential difference (voltage) from other forms of energy. The type of batteries you will use in these labs are known as chemical batteries because they convert internal chemical energy into electrical energy.

As a result of a potential difference, electric charge is repelled from one terminal of the battery and attracted to the other. However, no charge can flow out of a battery unless there is a conducting material connected between its terminals.

_____

*Some of the activities in this lab have been adapted from those designed by the Physics Education Group at the University of Washington.

If this conductor happens to be the filament in a small light bulb, the flow of charge will cause the light bulb to glow.

In this lab you are going to explore how charge flows in wires and bulbs when energy has been transferred to it by a battery. You will be asked to develop and explain some models that predict how the charge flows. You will also be asked to devise ways to test your models using current and voltage probes, which can measure the rate of flow of electric charge and the potential difference (voltage), respectively, and display these quantities on a computer screen.

INVESTIGATION 1: MODELS DESCRIBING CURRENT

What is electric current? The forces between objects that are rubbed in particular ways can be attributed to a property of matter known as *charge* (static electricity). Most textbooks assert that the electric currents through the wires connected to a battery are charges in motion. How do we know this? Perhaps current is something else—another phenomenon.

This is a question that received a great deal of attention from Michael Faraday, a famous early-nineteenth-century scientist. Faraday studied the effects of electricity from animals such as electric eels and tabulated his results in a table like the one shown in Figure 1-1. He concluded that "electricity, whatever may be its source, is identical in its nature."[1]

	Physiological effect	Magnetic deflection	Magnets made	Spark	Heating power	True chemical action	Attraction and repulsion	Discharge by hot air
1. Voltaic electricity	X	X	X	X	X	X	X	X
2. Common electricity	X	X	X	X	X	X	X	X
3. Magneto-electricity	X	X	X	X	X	X	X	
4. Thermo-electricity	X	X	+	+	+	+		
5. Animal electricity	X	X	X	+	+	X		

Figure 1-1: Reproduction of Faraday's table. The X's denote results obtained by Faraday and the +'s denote positive results found by other investigators later.

The purpose of the first activity is to compare carriers of the current produced by a battery to the static charges deposited by rubbing materials together. You will observe a demonstration using the following materials:

- 2 metal angle irons (approx. 15 cm long)
- foil-covered Styrofoam ball on a string (2.5 cm diameter)
- 300-V battery pack or power supply

[1]Faraday, M. "Identity of Electrocutes Derived from Different Sources," in *Experimental Researches in Electricity, Vol. I,* Taylor and Francis, London. (Reprinted by Dover Publications, New York, 1965, p. 76).

- rubber rod and cat fur

- glass rod and polyester cloth or silk

- electroscope

- alligator clip leads

- Wimshurst or Van de Graaff generator (optional)

Activity 1-1: Comparing Stuff from a Battery to the Rubbing Stuff

1. The angle irons, ball, and electroscope will be set up as shown in Figure 1-2.

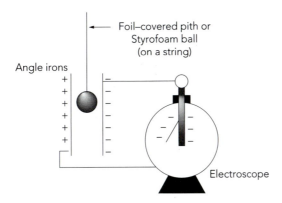

Figure 1-2: Apparatus for detecting charge.

2. The metal angle irons will be charged in two or more of the following ways:

   a. *Electrostatic Charging by Rubbing.* Stroke one plate with a rubber rod that has been rubbed with the cat fur. Repeat this several times. Stroke the other plate with the glass rod that has been rubbed with the polyester cloth or silk.

   b. *Charging with a Battery.* Connect a wire from the negative terminal of the battery pack to one of the angle irons. At the same time connect a wire from the positive terminal of the battery pack to the other plate.

   c. *Charging with a Wimshurst or Van de Graaff Generator.* Connect a wire from one of the two terminals of the generator to one angle iron, and a wire from the other terminal to the other angle iron.

3. Observe whether the different charging methods have different effects on the electroscope and on the ball when it is dangled between the two metal angle irons.

4. The metal angle irons will then be separated so the gap between them is just barely bigger than the diameter of the foil covered ball, and the ball will be carefully placed between them.

Question 1-1: What happened to the electroscope when the angle irons were charged with the familiar rubbing method—method (a)? Why?

Question 1-2: What happened when the ball was placed between the angle irons? In terms of the attraction and repulsion of different types of charges, explain why this unusual phenomenon happened.

Question 1-3: Describe what happened when the battery was used to "charge" the angle irons. What differences (if any) did you observe in the response of the electroscope and the ball to the charges on the angle irons?

Question 1-4: If the Wimshurst or Van de Graaff generator was also used, describe what happened. What differences (if any) did you observe in the response of the electroscope and the ball to the charges on the angle irons?

Question 1-5: Do the charges generated by rubbing and from the output of the battery cause different effects? If so, describe them. Do the charges generated in these two ways appear to be different?

The rate of flow of electric charge is more commonly called *electric current*. If charge Δq flows through the cross section of a conductor in time Δt, then the average current can be expressed mathematically by the relationship

$$\langle i \rangle = \frac{\Delta q}{\Delta t}$$

Instantaneous current is defined as the charge per unit time passing through a particular part of a circuit at an instant in time. It is usually defined using a limit:

$$i = \text{limit} \ \frac{\Delta q}{\Delta t}$$

as $\Delta t \rightarrow 0$

The unit of current is called the ampere (A). One ampere represents the flow of one coulomb of charge through a conductor in a time interval of one second. Another common unit is the milliampere (mA) (1 ampere = 1000 milliamperes). Usually people just refer to current as "amps" or "milliamps."

In the next activity, you can begin to explore electric current by lighting a bulb with a battery. You will need the following:

- flashlight bulb (#14)
- flashlight battery (1.5-V D cell)
- wire (6 inches or more in length)

Activity 1-2: Arrangements That Cause a Bulb to Light

Use the materials listed above to find some arrangements in which the bulb lights and some in which it does not light. For instance, try the arrangement shown in Figure 1-3.

Question 1-6: Sketch below two different arrangements in which the bulb lights.

Figure 1-3: A wiring configuration that might cause a bulb to light in the presence of a battery.

Question 1-7: Sketch below two arrangements in which the bulb *doesn't* light.

Question 1-8: Describe as fully as possible what conditions are needed if the bulb is to light, and why the bulb fails to light in the arrangements drawn in answer to Question 1-7.

Next you will explore which types of materials connected between the battery and the bulb allow the bulb to light and which do not. Since it seems that something flows from the battery to the bulb, we refer to materials that allow this flow as *conductors* and those that don't as *nonconductors*.
You will need

- common objects (paper clips, pencils, coins, rubber bands, fingers, paper, keys, etc.)

Activity 1-3: Other Materials Between the Battery and Bulb

Set up the single wire, battery, and bulb so that the bulb lights, e.g., one of the arrangements drawn in your answer to Question 1-6. Then, with the help of your partner, stick a variety of the common objects available between the battery and the bulb.

Question 1-9: List some materials that allow the bulb to light.

Question 1-10: List some materials that prevent the bulb from lighting.

Question 1-11: What types of materials seem to be conductors? What types seem to be nonconductors?

Are you having trouble holding your circuits together? Let's make it easier by using a battery holder and a bulb socket. While we're at it let's also add a switch in the circuit. In addition to the materials you've already used, you will need:

- battery holder (for a D cell)

- several wires (6 inches or more in length)

- flashlight bulb socket

- contact switch

Activity 1-4: Using a Battery Holder, Bulb Socket, and Switch

1. Examine the bulb socket carefully. Observe what happens when you unscrew the bulb.

2. Examine the bulb closely. Use a magnifying glass, if available. Figure 1-4 shows the parts of the bulb that are hidden from view.

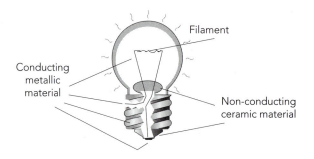

Figure 1-4: Wiring inside a light bulb.

Question 1-12: Why is the filament of the bulb connected in this way?

Question 1-13: Explain how the bulb socket works. Why doesn't the bulb light when it is unscrewed?

Prediction 1-1: If you wire up the configuration shown in Figure 1-5, will the bulb light with the switch open (i.e., so no contact between the wires is made)? Closed (i.e., so that contact is made)? Neither time? Explain your predictions.

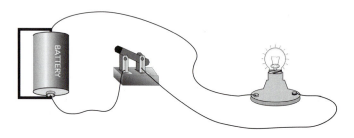

Figure 1-5: A circuit with a battery, switch, and bulb holder.

3. Wire the circuit shown in Figure 1-5 and test it.

4. Leave the switch closed so that the bulb remains on for 10–20 seconds. Feel the bulb.

Question 1-14: What did you feel? Besides giving off light, what happens to the bulb when there is a current through it?

Question 1-15: What do you conclude about the path needed by the current to make the filament heat up and the bulb glow? Explain based on all the observations you have made so far.

You are now ready to explore models for current in a circuit. The circuit to be considered is the one shown in Figure 1-5 with the switch closed. Figure 1-6 shows several models for the current in this circuit that are often proposed.

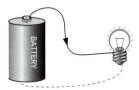

Model A
There will be no electric current left to flow in the bottom wire since the current is used up lighting the bulb.

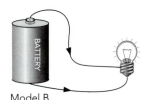

Model B
The electric current will travel in a direction toward the bulb in both wires.

Model C
The direction of the current will be in the direction shown, but there will be less current in the return wire since some of the current is used up lighting the bulb.

Model D
The direction of the current will be as shown, and the magnitude will be the same in both wires.

Figure 1-6: Four alternative models for current.

Prediction 1-2: Which model do you think best describes the current through the bulb? After you make your own prediction, talk things over with your partner. Explain your reasoning.

After you have discussed the various ideas with your partner, and chosen your favorite model, you can test your prediction. In Activity 1-5 you will use one or more current probes in your circuit to measure current. In addition to the battery, bulb, and wires you used above, you will need:

- computer-based laboratory system
- two current probes
- *RealTime Physics Electric Circuits* experiment configuration files

The current probe is a device that measures current and displays it as a function of time on the computer screen. It will allow you to explore the current at different locations and under different conditions in your electric circuits.

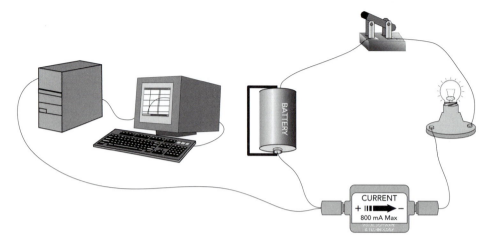

Figure 1-7: A circuit with a battery, bulb, switch, and current probe connected to the computer interface.

To measure the current through a part of the circuit, you must break open the circuit at the point where you want to measure the current, and insert the current probe. *That is, disconnect the circuit, put in the current probe, and reconnect with it in place.* For example, to measure the current in the bottom wire of the circuit in Figure 1-5, the current probe should be connected as shown in Figure 1-7.

Note that the current probe measures both the *magnitude* and *direction* of the current. A current that flows in through the + terminal and out through the − terminal (in the direction of the arrow) will be displayed as a positive current. Thus, if the current measured by the probe is positive, you know that the current must be counterclockwise in Figure 1-7 from the + terminal of the battery, through the bulb, through the switch, and toward the − terminal of the battery.

On the other hand, if the probe measures a negative current, then the current must be clockwise in Figure 1-7 (flowing into the − terminal and out of the + terminal of the probe).

Figure 1-8a shows a simplified diagram representing a current probe connected as shown in Figure 1-7.

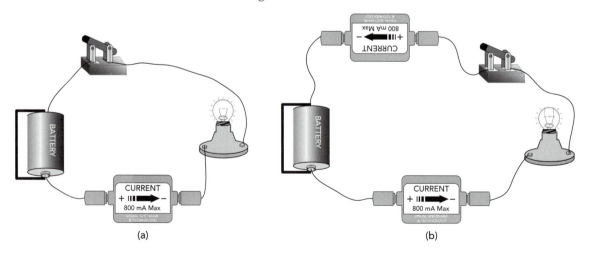

(a)   (b)

Figure 1-8 (a) Current probe connected to measure the current out of the + terminal of the battery and into the bulb. (b) Two current probes, one connected as in (a) and the other connected to measure the current out of the switch and into the − terminal of the battery.

Look at Figure 1-8b and convince yourself that if the currents measured by current probes 1 and 2 are both positive, this shows that the current is in a counterclockwise direction around all parts of the circuit.

You will use one or more bulbs and one or more current probes for the next activity. Design measurements that will allow you to choose the model (or models) that best describe the actual current through the circuit. (For example, to see if the current has a different magnitude or direction at different points in a circuit [model B or model C in Figure 1-6] you should connect two current probes in various locations around the circuit as in Figure 1-8b, to measure the current.)

Prediction 1-3: Use Table 1-1 to describe how the readings of current probe 1 and current probe 2 in the circuit in Figure 1-8b would compare with each other for each of the current models described in Figure 1-6.

Table 1-1

	Probe	Positive, negative, or zero?	CP1 > CP2, CP1 < CP2, or CP1 = CP2?
Model A	CP1		
	CP2		
Model B	CP1		
	CP2		
Model C	CP1		
	CP2		
Model D	CP1		
	CP2		

Activity 1-5: Developing a Model for Current in a Circuit

1. Be sure that current probes 1 and 2 are plugged into the interface.

2. Open the experiment file called **Current Model (L01A1-5).** The two sets of axes that follow should appear on the screen. The top axes display the current through probe 1 and the bottom the current through probe 2.

    The amount of current through each probe is also displayed digitally on the screen.

3. Be sure to **calibrate** the probes, or **load the calibration. Zero** the probes with them disconnected from the circuit.

4. To begin, set up the circuit in Figure 1-8b. **Begin graphing,** and try closing the switch for a couple of seconds and then opening it for a couple of seconds. Repeat this several times during the time when you are graphing. Sketch your graphs on the axes, or **print** your graphs and affix them over the axes.

Note: You should observe carefully whether the current through both probes is essentially the same or if there is a *significant* difference (more than a few percent).

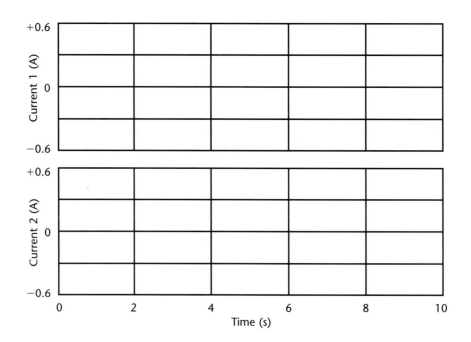

Question 1-16: How could you determine if an observed difference in the currents read by probe 1 and probe 2 is real or if it is the result of small calibration differences in the probes?

Question 1-17: Did you observe a *significant* difference in the currents at these two locations in the circuit, or was the current the same?

5. Try any other tests you need to decide on a current model among the choices in Figure 1-6, or any others you come up with. Sketch drawings of the circuits you used, showing where the probes were connected. **Print** all graphs, label them and affix them below.

Question 1-18: Based on your observations, which model seems to correctly describe the behavior of the current in your circuit. Explain carefully based on your observations.

Since you now know what it takes to get a light bulb to light, you can design and build some simple electrical devices before you learn more about what goes on in electric circuits. You can use extra switches, wires, bulbs, etc. as needed. You will have available

- 3 #14 bulbs and sockets
- 1.5-V D battery (must be very fresh alkaline) with holder
- 6 alligator clip leads
- single-pole–single-throw (SPST) switch
- single-pole–double-throw (SPDT) switch
- double-pole–double-throw (DPDT) switch

Activity 2-1: Inventing and Constructing Electric Circuits

Invent and construct one of the electric circuits described below. Sketch the circuit in the space below.

1. *Christmas Tree Lights:* Suppose you want to light up your Christmas tree with three bulbs. What happens if one of the bulbs fails? (Don't break the bulb! You can simulate failure by loosening a bulb in its socket.) Figure out a way to connect all three bulbs so that the other two will still be lit if any one of the bulbs burns out.

2. *Lighting a Tunnel:* The bulbs and switches must be arranged so that a person walking through a tunnel can turn on a lamp for the first part of the tunnel and then turn on a second lamp for the second half of the tunnel in such a way that the first one is turned off.

3. *Entry and Exit Light Switches:* A room has two doors. Light switches at both doors are wired so that either switch turns the lights in the room on and off.

Circuit diagram for circuit #____

Question 2-1: Describe how your circuit works, and why you connected it in the way you did.

Extension 2-2: Inventing other Electric Circuits

Invent and construct one or both of the other electric circuits described in Activity 2-1, or invent your own circuit that performs a certain task. Sketch each circuit diagram and describe how the circuit works in the space below.

Now that you have been wiring circuits and drawing diagrams of them you may be getting tired of drawing pictures of the batteries, bulbs, and switches. There are a series of symbols that have been created to represent circuits. A few of the electric circuit symbols are shown in Figure 1-9.

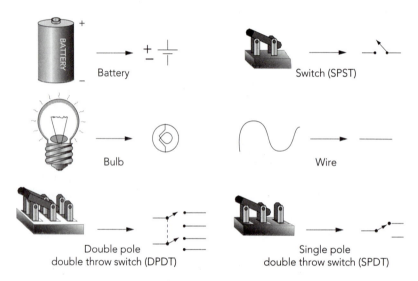

Figure 1-9: Some common circuit symbols.

Using these symbols, the circuit from Figure 1-5 with a switch, bulb, wires, and battery can be sketched as on the right in Figure 1-10.

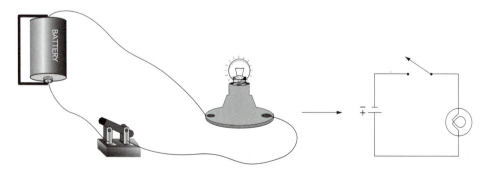

Figure 1-10: A circuit sketch and corresponding circuit diagram.

REALTIME PHYSICS: ACTIVE LEARNING LABORATORY

Activity 2-3: Drawing Circuit Diagrams

Sketch a nice neat "textbook" style circuit diagram for each of the circuits you designed in Activity 2-1 and Extension 2-2.

Question 2-2: On the battery symbol, which line represents the positive terminal—the long one or the short one? [**Note:** You should try to remember this convention for the battery polarity because some circuit elements, such as *diodes*, behave differently if the battery is turned around so it has opposite polarity.]

There are actually two important quantities to consider in describing the operation of electric circuits. One is *current*, and the other is *potential difference*, often referred to as *voltage*. In Activity 1-5 you measured the current at two different positions in a circuit. Now, as an introduction to our studies of more complex circuits, let's actually measure *both current and voltage* in a familiar circuit.

In addition to the equipment you have been using so far, you will need:

- computer-based laboratory system
- two voltage probes
- two current probes
- *RealTime Physics Electric Circuits* experiment configuration files

Figure 1-11 shows the symbols we will use to indicate a current probe or a voltage probe.

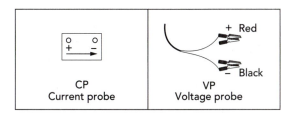

Figure 1-11: Symbols for current probe and voltage probe.

Figure 1-12a shows our simple circuit from Figure 1-5 with voltage probes connected to measure the voltage *across* the battery and the voltage *across* the bulb. The circuit is drawn again symbolically in Figure 1-12b. Note that the word *across* is very descriptive of how the voltage probes are connected.

Activity 2-4: Measuring Potential Difference (Voltage) and Current

1. To set up the voltage probes, first unplug the current probes from the interface and plug in the voltage probes.

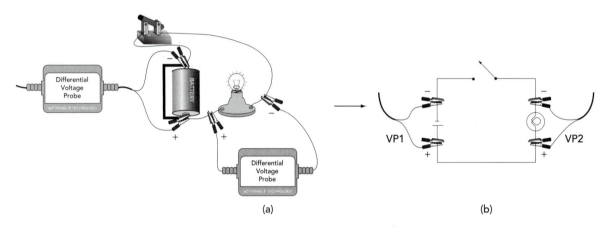

(a)   (b)

Figure 1-12: Two voltage probes connected to measure the voltages across the battery and the bulb.

2. Open the experiment file called **Two Voltages (L01A2-4a)** to display the axes for two voltage probes that follow.

3. **Zero** the voltage probes with them disconnected from the circuit.

4. Connect the circuit shown in Figure 1-12, but *do not connect the probes yet*.

5. First connect both the + and the − clips of one voltage probe together. Observe the reading. Next connect both clips to the *same point* in the circuit. Close the switch.

6. Finally connect the + clip to the + end of the battery and the − clip to the side of the bulb indicated with a + in Figure 1-12. Close the switch.

Question 2-3: What do you conclude about the voltage when the voltage probe leads are connected to the same point or to the two ends of the same wire?

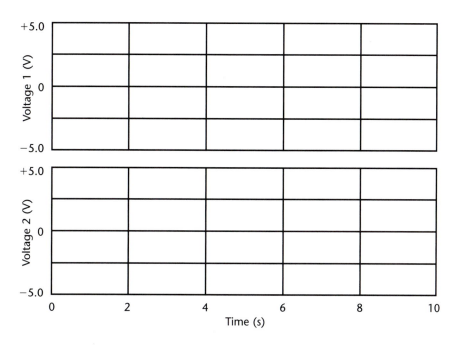

Prediction 2-1: In the circuit in Figure 1-12, how would you expect the voltage across the battery to compare to the voltage across the bulb with the switch open and closed? Explain.

7. Now test your prediction. Connect the voltage probes to measure the voltage *across* the battery and the voltage *across* the bulb simultaneously.

8. **Begin graphing,** and close and open the switch several times. **Print** your graphs and affix them over the axes above, or sketch them on the axes.

Question 2-4: What do you conclude about the voltage across the battery and the voltage across the bulb when the switch is open and when it is closed? Are the graphs the same? Why or why not? What is going on as the switch is closed and opened?

9. Now connect a voltage and a current probe so that you are measuring the voltage *across* the battery and the current *through* the battery at the same time (see Figure 1-13).

10. Open the experiment file called **Current and Voltage (L01A2-4b)** to display the current and voltage axes that follow.

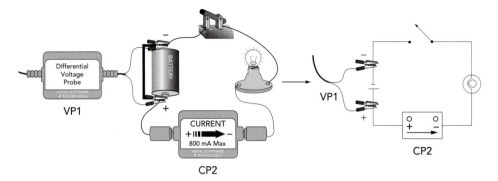

Figure 1-13: Probes connected to measure the voltage across the battery and the current through it.

11. **Begin graphing,** and close and open the switch several times, as before. Sketch your graphs, or **print** them and affix over the axes.

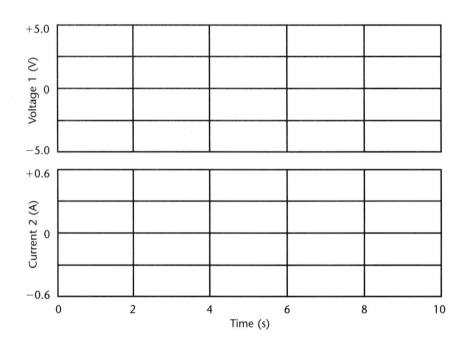

Question 2-5: Explain the appearance of your current and voltage graphs. What happens to the current through the battery as the switch is closed and opened? What happens to the voltage across the battery?

12. Find the voltage across and the current through the battery when the switch is closed and the bulb is lit. (You can use the digital display on the computer screen.)

Average voltage: ___   Average current: ___

Prediction 2-2: Now suppose you connect a second bulb in the circuit, as shown in Figure 1-14. How do you think the voltage across the battery will compare to that with only one bulb? Will it change significantly? Explain.

13. Connect the circuit with two bulbs and test your prediction. Again measure the voltage across and the current through the battery with the switch closed.

Average voltage: ____   Average current: ____

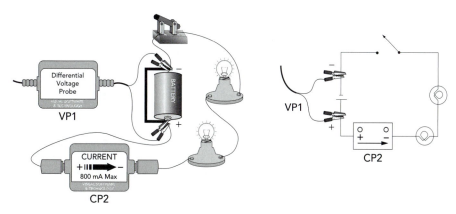

Figure 1-14: Two bulbs connected with voltage and current probes.

Question 2-6: Did the current through the battery change significantly when you added the second bulb to the circuit (by more than several percent)?

Question 2-7: Did the voltage across the battery change significantly when you added the second bulb to the circuit (by more than several percent)?

Question 2-8: Does the battery appear to be a source of constant current, constant voltage, or neither when different elements are added to a circuit?

INVESTIGATION 3: AN ANALOGY TO CURRENT AND RESISTANCE

You found in Activity 1-5 that current is not used up in passing through a bulb, but this may seem counterintuitive to you. Also, how can we explain that there is less current in the circuit with two bulbs instead of one? Lots of physics teachers have invented analogies to help explain these electric circuit concepts. One approach is to construct a model of a gravitational system that is in some ways analogous to the electrical system you are studying (Figure 1-5).

It is believed that the electrons flowing through a conductor have frequent collisions that slow them down and change their directions. Between collisions each electron accelerates and finally staggers through the material with an average drift velocity, $\langle \vec{v}_{\text{drift}} \rangle$.

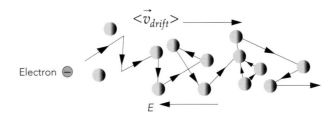

Figure 1-15: A simplified depiction of an electron in a uniform electric field staggering through a conductor as a result of collisions. Despite the constant force to the right caused by the electric field, these collisions cause the electron to move through the conductor with a constant average velocity, $\langle \vec{v}_{\text{drift}} \rangle$, the "drift velocity" (instead of accelerating).

As you saw in Investigation 2, we can talk about the *resistance* to flow of electrons that materials offer. A wire has a low resistance. A light bulb has a much higher resistance. Special electric elements that resist current are called *resistors*. You will examine the behavior of these in electric circuits in future labs.

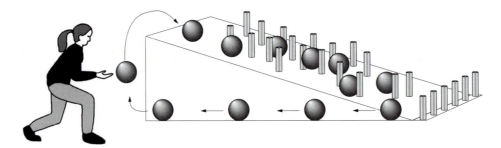

Figure 1-16: An analog to electric current and resistance.

It is possible to use a two-dimensional mechanical analog to model this picture of current through conductors. You should note that the real flow of electrons is a three-dimensional affair. The diagram for the two-dimensional analog is reproduced in Figure 1-16.

Extension 3-1: Drawing an Analogy

Look at the current and resistance analog in Figure 1-16, and then answer the following questions.

Question E3-1: What part of the picture represents the action of the battery? What represents the electric charge and current? What part represents the collisions of electrons? Explain.

Question E3-2: What ultimately happens to the "energy" given to the bowling balls by the "battery"? What plays the role of the bulb? How is this energy loss exhibited in the circuit you wired that consists of a battery, two wires, and a bulb?

Question E3-3: How does this model help explain the fact that electric current doesn't decrease as it passes through the bulb?

Question E3-4: How does this model help explain the fact that electrons move with a constant average speed v_{drift}, rather than having a constant acceleration caused by the constant electric field?

Question E3-5: In this model what would happen to the "ball" current if the drift velocity doubled? What can you do to the ramp to increase the drift velocity?

HOMEWORK FOR LAB 1
BATTERIES, BULBS, AND CURRENT

1. Is there any difference between the static charges generated by rubbing a glass rod with silk or a rubber rod with cat fur and the charges that flow (from a battery) through wires in an electric circuit? Give evidence for your answer.

2. For the circuit on the right, indicate whether the statements below are TRUE or FALSE. If a statement is TRUE, briefly describe the evidence from this lab which supports this statement. If a statement is FALSE, give a correct statement, and briefly describe the evidence from this lab which supports this new statement.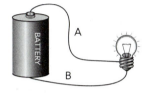

    a. The current is from the battery, through wire A, through the bulb, and then back to the battery through wire B.

    b. Since current is used up by the bulb, the current in wire B is smaller than the current in wire A.

    c. The current is toward the bulb in both wires A and B.

    d. If wire B is disconnected, but wire A is left connected, the bulb will still light, but if wire A is disconnected and wire B is left connected, the bulb will not light.

    e. A current probe will read the same magnitude if connected to measure the current in wire A or wire B.

3. What circuit element is represented by each of the following symbols?

    a. ─┴─ b. ─•╲── c. ──── d. ──⊙──

4. Draw below a circuit diagram using the symbols in Question 3 for at least one circuit you worked on in Activity 2-1.

5. Draw below a circuit diagram for the circuit in Question 2 with one current probe hooked up to measure the current in wire A and a voltage probe hooked up to measure the voltage across the light bulb. Also include a switch in the circuit to turn the bulb on and off. Use correct symbols for all circuit elements.

6. Consider the two circuits below. All bulbs and all batteries are identical. Compare the voltage across the battery in the left circuit to that in the right circuit. *Describe the evidence in this lab for your answer.*

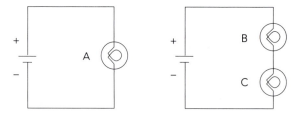

Pre-Lab Preparation Sheet for Lab 2—Current in Simple DC Circuits

(Due at beginning of lab)

Directions:
Read over Lab 2 and then answer the following questions about the procedures.

1. What do you predict for the rankings of the brightness of bulbs A, B, and C in Figure 2-1?

2. How do you predict that changing the direction of the current by reversing the connections to the battery in Figure 2-1 would change the rankings in (1)?

3. How will you compare the currents in the circuits in Figure 2-1 experimentally? What equipment will you use?

4. Define *series* and *parallel* connections. Sketch two light bulbs connected in series and to a battery, and two light bulbs connected in parallel and to a battery.

5. How do you predict the brightness of bulb D will change when the switch is closed in Figure 2-6?

6. How do you predict the current through the battery will change when the switch is closed in Figure 2-6?

LAB 2:
CURRENT IN SIMPLE DC CIRCUITS*

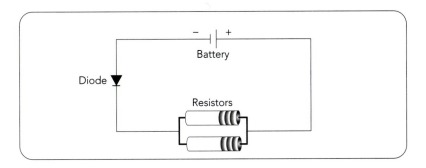

If it's green and it wiggles, it's biology.

If it stinks, it's chemistry.

If it doesn't work, it's physics.

If it's incomprehensible, it's mathematics.

If it doesn't make sense, it's either economics or psychology.

—From A. Bloch's
Murphy's Law Book 3

OBJECTIVES

- To understand current in a circuit where a battery lights a bulb.

- To understand the meaning of series connections in an electric circuit.

- To understand the relationship between the current in all parts of a series circuit.

- To understand the meaning of parallel connections in an electric circuit.

- To understand the relationship between the currents in all parts of a parallel circuit.

- To begin to understand the concept of resistance.

OVERVIEW

In the Lab 1 you saw that when there is an electric current through a light bulb, the bulb lights. You also saw that to cause current through a bulb, you must connect the bulb in a complete circuit with a battery. There will be a current only when there is a complete path from the positive terminal of the battery, through the connecting wire to the bulb, through the bulb, through the connecting wire to the negative terminal of the battery, and through the battery.

By measuring the current at different points in a simple circuit consisting of a bulb, a battery, and connecting wires, you discovered a model for current, namely

---

*Some of the activities in this lab have been adapted from those designed by the Physics Education Group at the University of Washington.

that the electric current was the same in all parts of the circuit. By measuring the current and voltage in this circuit and adding a second bulb, you also discovered that a battery maintains essentially the same voltage whether it is connected to one light bulb or two.

You also observed that the current was smaller when a second bulb was added to the circuit. This led us to introduce the concept of *resistance* of a circuit element such as a bulb. The total resistance of a circuit determines the current when the circuit is connected to a battery.

In this lab you will examine more complicated circuits than a single bulb connected to a single battery. You will compare the currents through different parts of these circuits by comparing the brightness of the bulbs, and also by measuring the currents using current probes. In Lab 3, you will further examine the role of the battery in causing a current in a circuit, and compare the potential differences (voltages) across different parts of your circuits.

INVESTIGATION 1: CURRENT IN SERIES CIRCUITS

In the next series of activities you will be asked to make a number of predictions about the current in various circuits and then to compare your predictions with actual observations. Whenever your experimental observations disagree with your predictions you should try to develop new concepts about how circuits with batteries and bulbs actually work. To make the required observations you will need the following items:

- computer-based laboratory system
- two current probes
- *RealTime Physics Electric Circuits* experiment configuration files
- 1.5-V D battery (must be very fresh, alkaline) with holder
- 6 wires with alligator clip leads
- 4 #14 bulbs in sockets
- contact switch

Prediction 1-1: Consider the two circuits shown in Figure 2-1.

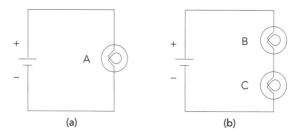

(a)   (b)

Figure 2-1: Two different circuits: (a) a battery with a single bulb, and (b) a battery with two identical bulbs identical to the one in (a).

Predict the relative brightness of the three bulbs shown in Figure 2-1 from brightest to dimmest, assuming that the batteries and bulbs have identical characteristics. (Remember that you saw in the last laboratory that the battery maintains essentially the same *voltage* across its terminals whether there is one light bulb or two.)

If two or more bulbs are equal in brightness, indicate this in your response. Explain the reasons for your rankings.

Hint: Helpful symbols are >, "is greater than"; <, "is less than"; =, "is equal to." For example, B > C > A.

Activity 1-1: The Relative Brightness of Bulbs

1. Now connect the circuits, observe the relative brightness of the bulbs, and rank the bulbs in order of the brightness you actually observed.

Comment: These activities assume *identical* bulbs. Differences in brightness may arise if the bulbs are not exactly identical. In this and later activities, to determine whether a difference in brightness is caused by a difference in the currents through the bulbs or by a difference in the bulbs, you should exchange the bulbs.

Sometimes a bulb will not light noticeably, even if there is a small but significant current through it. If a bulb is really off, that is, if there is no current through it, then unscrewing the bulb will not affect the rest of the circuit. To verify whether a nonglowing bulb actually has a current through it, unscrew the bulb and see if anything else in the circuit changes.

Ranking of the bulbs:

Question 1-1: Did your observations agree with your predictions? If not, explain what assumptions you were making that now seem false.

Prediction 1-2: What do you predict will happen to the brightness of bulbs A, B, and C in Figure 2-1 if the battery is connected to the bulbs with its terminals reversed? Explain the reason(s) for your prediction.

2. Test your prediction. Reverse the terminals of the battery in each of your circuits.

Question 1-2: What do you observe? Does this agree with your prediction? Did you make any false assumptions? Explain.

Question 1-3: Can you tell anything about the direction of the current through the circuit by just looking at the brightness of the bulbs without knowing how the battery is hooked up? Explain.

How does each bulb affect the current in a circuit? Does current get used up after passing through a bulb? How does the current in the two-bulb circuit compare to that in the single-bulb circuit? First make predictions and then observe experimentally.

Prediction 1-3: What would you predict about the relative amount of current going through each bulb in Figures 2-1a and b? Write down your predicted rankings of the currents through bulbs A, B, and C, and explain your reasoning.

Activity 1-2: Current in a Simple Circuit with Bulbs

You can test your prediction by using current probes. Recall from Lab 1 that to measure the current through a bulb, a current probe must be connected so that the current through the current probe is the same as the current through the bulb. Convince yourself that the current probes shown in Figure 2-2 measure the currents described in the figure caption.

> **Comment:** In carrying out your measurements, it is important to realize that the measurements made by the current probes are only as good as their calibrations. Small differences in calibration can result in small differences in readings.

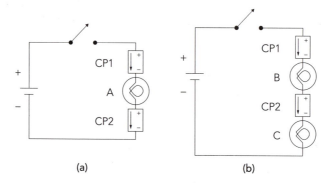

(a)   (b)

Figure 2-2: Current probes connected to measure the current through bulbs. In circuit (a), CP1 measures the current into bulb A, and CP2 measures the current out of bulb A. In circuit (b), CP1 measures the current into bulb B while CP2 measures the current out of bulb B and the current into bulb C.

1. Open the experiment **Two Currents (L02A1-2)** to display the two sets of current axes that follow.

2. **Calibrate** the current probes, or load the calibration. **Zero** the probes with them disconnected from the circuit.

3. Connect circuit (a) in Figure 2-2.

4. **Begin graphing,** close the switch for a second or so, open it for a second or so, and then close it again. Sketch your graphs on the axes, or **print** your graphs and affix them over the axes.

5. Use the **analysis feature** of the software to measure the currents into and out of bulb A when the switch is closed:

Current into bulb A:_____   Current out of bulb A:_____

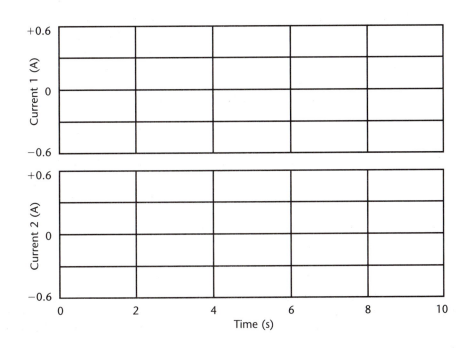

Question 1-4: Are the currents into and out of bulb A equal or is one significantly larger (do they differ by more than a few percent)? What can you say about the directions of the currents? Is this what you expected? Explain.

6. Connect circuit (b) in Figure 2-2. **Begin graphing** current as above, and record the measured values of the currents.

Current through bulb B:_____Current through bulb C:_____

7. Sketch the graphs on the axes that follow, or **print** and affix over the axes.

Question 1-5: Consider your observation of the circuit with bulbs B and C in it. Is current "used up" in the first bulb or is it the same in both bulbs? Explain based on your observations.

Question 1-6: Is the ranking of the currents in bulbs A, B, and C what you predicted? If not, can you explain what assumptions you were making that now seem false?

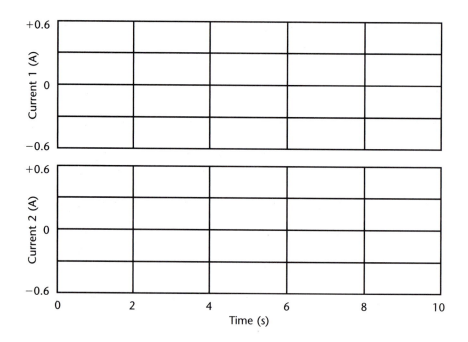

Question 1-7: Based on your observations, how is the brightness of a bulb related to the current through it?

Question 1-8: How does the amount of current produced by the battery in the single bulb circuit (Figure 2-1a) compare to that produced by the battery with two bulbs connected as in Figure 2-1b? Does the addition of this second bulb in this manner affect the current through the original bulb? Explain.

Question 1-9: Suppose you think of the bulb as providing a *resistance* to the current in a circuit, rather than something that uses up current. How do you expect the total resistance in a circuit is affected by the addition of more bulbs in the manner shown in Figure 2-1b?

Question 1-10: Formulate a rule for predicting whether current increases or decreases as the total resistance of the circuit is increased.

Comment: The rule you have formulated based on your observations with bulbs may be *qualitatively* correct—correctly predicting an increase or decrease in current—but it won't be *quantitatively* correct. That is, it won't allow you to predict the exact sizes of the currents correctly. This is because the resistance of a bulb to current changes as the current through the bulb changes. You will explore this in more detail in Lab 3.

Another common circuit element is a *resistor*. A resistor has a constant resistance to current regardless of how large the current through it. In the next activity you will reformulate your rule using resistors.

First a prediction.

Prediction 1-4: Consider the circuit diagrams in Figure 2-3 in which the light bulbs in Figure 2-1 have been replaced by identical resistors.

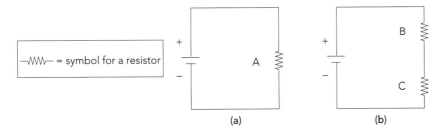

Figure 2-3: Two different circuits: (a) a battery with a single resistor, and (b) a battery with two resistors identical to the one in (a).

What would you predict about the relative amount of current going through each resistor in Figures 2-3a and b? Write down your predicted rankings of the currents through resistors A, B, and C, and explain your reasoning. (Remember that a resistor has a constant resistance to current regardless of the current through it.)

In addition to the equipment listed above, you will need the following to test your predictions:

- two 10-Ω resistors

Activity 1-3: Current in a Simple Circuit with Resistors

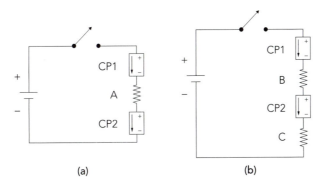

Figure 2-4: Current probes connected to measure the current through resistors. In circuit (a), CP1 measures the current into resistor A, and CP2 measures the current out of resistor A. In circuit (b), CP1 measures the current into resistor B, while CP2 measures the current out of resistor B and into resistor C.

1. Continue to use the experiment file **Two Currents (L02A1-2)**.

2. **Calibrate** the current probes, or load the calibration if this has not already been done. **Zero** the probes with them disconnected from the circuit.

3. Connect circuit (a) in Figure 2-4.

4. Use the current probes and the **analysis feature** in the software to measure the current through resistor A in circuit 2-4a and the currents through resistors B and C in circuit 2-4b.

   Current through resistor A:_____

   Current through resistor B:_____   Current through resistor C:_____

Question 1-11: Is the ranking of the currents in resistors A, B, and C what you predicted? If not, can you explain what assumptions you were making that now seem false?

Question 1-12: How does the amount of current produced by the battery in the single resistor circuit (Figure 2-3a) compare to that produced by the battery with two resistors connected as in Figure 2-3b? Does the addition of this second resistor in this manner affect the current through the original resistor? Explain.

Question 1-13: How did your observations with resistors differ from your observations with bulbs in Activity 1-2.

Question 1-14: Reformulate a more quantitative rule for predicting *how* the current supplied by the battery decreases as more *resistors* are connected in the circuit as in Figure 2-3b.

Question 1-15: Is your rule in Question 1-14 also correct for bulbs connected as in Figures 2-1a and b? Explain.

INVESTIGATION 2: CURRENT IN PARALLEL CIRCUITS

There are two basic ways to connect resistors, bulbs, or other elements in a circuit: *series* and *parallel*. So far you have been connecting bulbs and resistors *in series*. To make predictions involving more complicated circuits we need to have a more precise definition of series and parallel. These are summarized below.

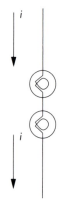

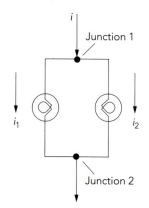

Series connection:
Two resistors are in series if they are connected so that the same current that passes through one bulb passes through the other.

Parallel connection:
Two resistors are in parallel if their terminals are connected so that at each junction one terminal of one bulb is directly connected to one terminal of the other.

It is important to keep in mind that in more complex circuits, say with three or more elements, not every element is necessarily connected in series or parallel with other elements.

Let's compare the behavior of a circuit with two bulbs wired in parallel to the circuit with a single bulb (see Figure 2-5).

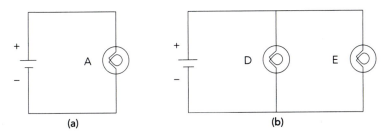

(a)   (b)

Figure 2-5: Two different circuits: (a) a single-bulb circuit and (b) a circuit with two bulbs identical to the one in (a) connected *in parallel* to each other and *in parallel* to the battery.

Note that if bulbs A, D, and E are identical, then the circuit in Figure 2-6 is equivalent to circuit 2-5a when the switch is open (as shown) and equivalent to circuit 2-5b when the switch is closed.

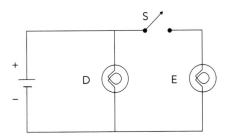

Figure 2-6: When the switch is open, only bulb D is connected to the battery. When the switch is closed, bulbs D and E are connected *in parallel* to each other and *in parallel* to the battery.

Question 2-1: Explain how you know that the caption of Figure 2-6 correctly describes the circuit.

Prediction 2-1: Predict how the brightness of bulbs D and E in the parallel circuit of Figure 2-5b will compare to bulb A in the single bulb circuit of Figure 2-5a. How will D and E compare with each other? Rank the brightness of all three bulbs. Explain the reasons for your predictions.

To test this and other predictions, you will need:

- computer-based laboratory system
- two current probes
- *RealTime Physics Electric Circuits* experiment configuration files

- 1.5-V D battery (must be very fresh, alkaline) with holder
- 8 wires with alligator clip leads
- 3 #14 bulbs in sockets
- contact switch

Activity 2-1: Brightness of Bulbs in a Parallel Circuit

Set up the circuit in Figure 2-6, and describe your observed rankings for the brightness of bulb D with the switch open, and D and E with the switch closed.

Question 2-2: Did the observed rankings agree with your prediction? If not, can you explain what assumptions you were making that now seem false?

Prediction 2-2: Based on your observations of brightness, what do you predict about the relative amount of current through each bulb in a parallel connection, i.e., bulbs D and E in Figure 2-5b?

Prediction 2-3: Based on your observations of brightness, how do you think that closing the switch in Figure 2-6 affects the current through bulb D?

Activity 2-2: Current in Parallel Branches

You can test Predictions 2-2 and 2-3 by connecting current probes to measure the currents through bulbs D and E.

1. Open the experiment file called **Two Currents (L02A1-2),** if it is not already opened.

2. **Calibrate** the current probes, or load the calibration, if this hasn't already been done. **Zero** the probes with them disconnected from the circuit.

3. Connect the circuit shown in Figure 2-7.

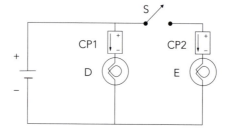

Figure 2-7: Current probes connected to measure the current through bulb D and the current through bulb E.

4. **Begin graphing** the currents through both probes then close the switch for a second or so, open it for a second or so, and then close it again.

5. Sketch the graphs on the axes below, or **print** them and affix them over the axes.

6. Use the **analysis feature** of the software to measure both currents.

    Switch open: Current through bulb D:_____ Current through bulb E:_____

    Switch closed: Current through bulb D:_____ Current through bulb E:_____

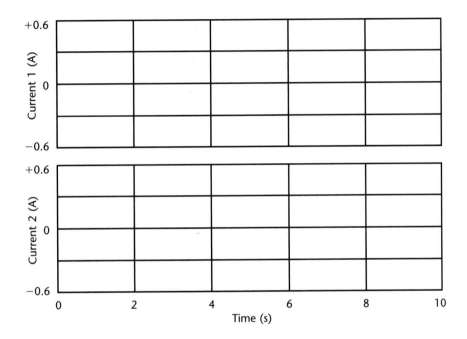

Question 2-3: Based on your graphs and measurements, were the currents through bulbs D and E what you predicted based on their brightness? If not, can you now explain why your prediction was incorrect?

Question 2-4: Did closing the switch and connecting bulb E *in parallel* with bulb D significantly affect the current through bulb D? How do you know? [**Note:** You are making a *very significant* change in the circuit. Think about whether the new current through D when the switch is closed reflects this.]

You have already seen in Lab 1 that the voltage maintained by a battery doesn't change appreciably no matter what is connected to it. But what about the current through the battery? Is it always the same no matter what is connected to it, or does it change depending on the circuit? (Is the current through the battery the same whether the switch in Figure 2-6 is open or closed?) This is what you will investigate next.

Prediction 2-4: Based on your observations of the brightness of bulbs D and E in Activity 2-2, what do you predict about the amount of current through the battery in the parallel bulb circuit (Figure 2-5b) compared to that through the single bulb circuit (Figure 2-5a)? Explain.

Activity 2-3: Current Through the Battery

1. Test your prediction with the circuit shown in Figure 2-8. Use the same experiment file, **Two Currents (L02A1-2),** as in the previous activities.

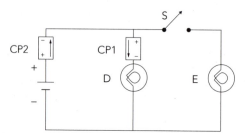

Figure 2-8: Current probes connected to measure the current through the battery and the current through bulb D.

2. **Begin graphing** while closing and opening the switch as before. Sketch your graphs on the axes that follow, or **print** and affix over the axes. Label on your graphs when the switch is open and when it is closed.

3. Measure the currents through the battery and through bulb D.

   Switch open: Current through battery:_____ Current through bulb D:____

   Switch closed: Current through battery:_____ Current through bulb D:____

Question 2-5: Describe how the connection of current probes in Figure 2-8 differs from that in Figure 2-7. How do you know that probe 2 is measuring the current through the battery?

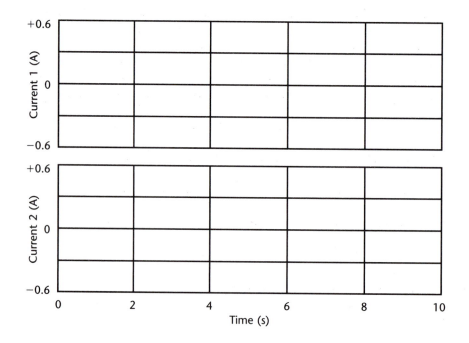

Question 2-6: Use your observations to formulate a rule to predict how the current through a battery will change as the number of bulbs connected *in parallel* increases. Can you explain why?

Question 2-7: Comparing your rule in Question 2-6 to the rule you stated in Questions 1-10 and 1-14 relating the current through the battery to the total *resistance* of the circuit, does the addition of more bulbs in parallel increase, decrease, or not change the total *resistance* of the circuit? Explain.

Question 2-8: Can you explain your answer to Question 2-7 in terms of the number of paths for current available in the circuit? Explain.

Question 2-9: Considering your experiences with series and parallel circuits in Investigations 1 and 2, does the current through the battery depend only on the number of bulbs or resistors in the circuit, or does the arrangement of the circuit elements matter?

Question 2-10: Since current and resistance are related, does the resistance depend just on the number of bulbs or resistors, or does it depend on the arrangement of the circuit elements as well? Explain.

INVESTIGATION 3: MORE COMPLEX SERIES AND PARALLEL CIRCUITS

Now you can apply your knowledge to some more complex circuits. Consider the circuit consisting of a battery and two bulbs, A and B, in series shown in Figure 2-9a. What will happen if you add a third bulb, C, in parallel with bulb B as shown in Figure 2-9b? You should be able to predict the relative brightness of A, B, and C based on previous observations. The tough question is: how does the brightness of A change when C is connected in parallel to B?

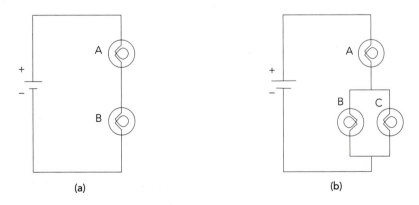

Figure 2-9: Two circuits with identical batteries and bulbs A, B, and C.

Question 3-1: In Figure 2-9b is bulb A in series with bulb B? with bulb C? or with a combination of bulbs B and C? (You may want to go back to the definitions of series and parallel connections at the beginning of Investigation 2.)

Question 3-2: In Figure 2-9b are bulbs B and C connected in series or in parallel with each other, or neither? Explain.

Question 3-3: Is the resistance of the combination of bulbs B and C larger than, smaller than, or the same as bulb B alone? Explain.

Question 3-4: Is the resistance of the combination A, B, and C in Figure 2-9b larger than, smaller than, or the same as the combination of A and B in Figure 2-9a? Explain.

Prediction 3-1: Predict how the current through bulb A will change, if at all, when circuit 2-9a is changed to 2-9b (when bulb C is added in parallel to bulb B). What will happen to the brightness of bulb A? Explain the reasons for your predictions.

Prediction 3-2: Predict how the current through bulb B will change, if at all, when circuit 2-9a is changed to 2-9b (when bulb C is added in parallel to bulb B). What will happen to the brightness of bulb B? Explain the reasons for your predictions.

Prediction 3-3: Also predict the relative rankings of brightness for all the bulbs, A, B, and C, after bulb C is in the circuit. Explain the reasons for your predictions.

Activity 3-1: A More Complex Circuit

1. Set up the circuit shown in Figure 2-10a. Convince yourself that this circuit is identical to Figure 2-9a when the switch, S, is open, and to Figure 2-9b when the switch is closed.

2. Observe the brightness of bulbs A and B when the switch is open, and then the brightness of the three bulbs when the switch is closed. Compare the brightness of bulb A with the switch open and closed, and rank the brightness of bulbs A, B, and C with the switch closed.

Question 3-5: If you did not observe what you predicted about the brightness, what changes do you need to make in your reasoning? Explain.

3. Connect the two current probes as shown in Figure 2-10b. Open the experiment file called **Two Currents (L02A1-2),** if it is not already opened.

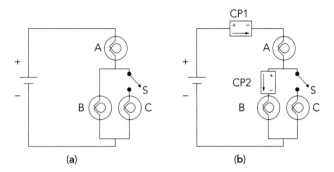

(a)   (b)

Figure 2-10: (a) Circuit equivalent to Figure 2-9a when the switch, S, is open and to Figure 2-9b when the switch is closed. (b) Same circuit with current probes connected to measure the current through bulb A (CP1) and the current through bulb B (CP2).

4. **Begin graphing** and observe what happens to the current through bulb A (through the battery) and the current through bulb B when the switch is opened and closed.

Question 3-6: What happens to the current through the battery and through bulbs A and B when bulb C is added in parallel with bulb B? What do you conclude happens to the total resistance in the circuit? Explain.

If you have additional time, do some or all of the following Extensions to examine some more complex circuits.

Extension 3-2: An Even More Complex Circuit

Let's look at a somewhat more complicated circuit to see how series and parallel parts of a complex circuit affect one another. The circuit is shown in Figure 2-11.

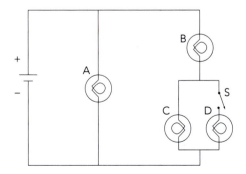

Figure 2-11: A complex circuit with series and parallel connections.

Question E3-7: When switch S is open, which bulbs are connected in parallel with each other? (If you need to, review the definitions of series and parallel at the beginning of Investigation 2 before answering.)

Is A parallel to B?

Is A parallel to C?

Is C parallel to D?

Is A parallel to the combination of B and C?

Question E3-8: When switch S is open, which bulbs are connected in series with each other?

Is A in series with B?

Is A in series with C?

Is B in series with C?

Question E3-9: When switch S is closed, which bulb(s) are connected in parallel with A?

Question E3-10: When switch S is closed, which bulb(s) are connected in series with B?

Prediction E3-4: Predict the effect on the current through bulb A for each of the following separate alterations in the circuit:

a. unscrewing bulb B

b. closing switch S

Prediction E3-5: Predict the effect on the current through bulb B of each of the following separate alterations in the circuit:

a. unscrewing bulb A

b. adding another bulb in series with bulb A

Connect the circuit in Figure 2-11, and observe the effect of each of the alterations in Predictions E3-4 and E3-5 on the brightness of each bulb. Describe your observations.

Question E3-11: Compare your results with your predictions. How do you account for any differences between your predictions and observations?

Question E3-12: In this circuit, two parallel branches are connected *directly across* a battery. For this type of connection, what do you conclude about the effect of changes in one parallel branch on the current in the other?

Extension 3-3: Series and Parallel Networks

Now let's practice with some more complicated series and parallel circuits. Suppose you had three boxes, labeled A, B, and C, each having two terminals. The arrangement of resistors in the boxes is shown in Figure 2-12.

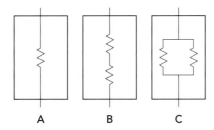

A   B   C

Figure 2-12: Parallel and series circuits.

Consider the five circuits shown in Figure 2-13 in completing the next activity.

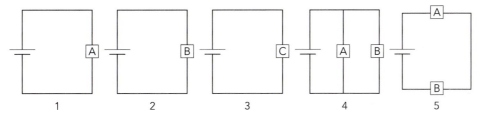

1   2   3   4   5

Figure 2-13: Circuits featuring parallel and series connections.

Question E3-13: For each of the circuits in Figure 2-13, sketch a standard circuit diagram showing all the resistors in the circuit. In each diagram number the resistors and describe which resistors or combination of resistors are in series and parallel with each other.

Question E3-14: Rank the five circuits in Figure 2-13 by their total resistance. Which has the most resistance? The least resistance? Explain your reasoning.

Question E3-15: Rank each of the circuits in Figure 2-13 according to the total current through the battery. Explain your reasoning.

You can now test your understanding of current and resistance on another puzzling circuit.

Extension 3-4: The Puzzle Problem

Question E3-16: Use reasoning based on your model of electric current to predict the relative brightness of each of the bulbs shown in Figure 2-14. Explain the reasons for your rankings.

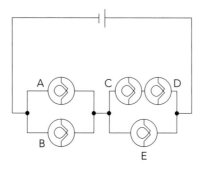

Figure 2-14: A complicated battery and bulb circuit.

HOMEWORK FOR LAB 2
CURRENT IN SIMPLE DIRECT CURRENT CIRCUITS

1. Which of the three circuits shown below, if any, are the same electrically? Which are different? Explain your answers.

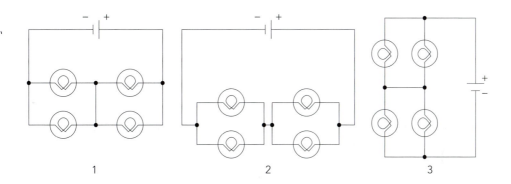

2. Consider the two messy circuit diagrams 1 and 2 below.

   a. Identify which of the nice neat circuit diagrams below (A, B, C, or D) corresponds to circuit 1. Explain the reasons for your answer.

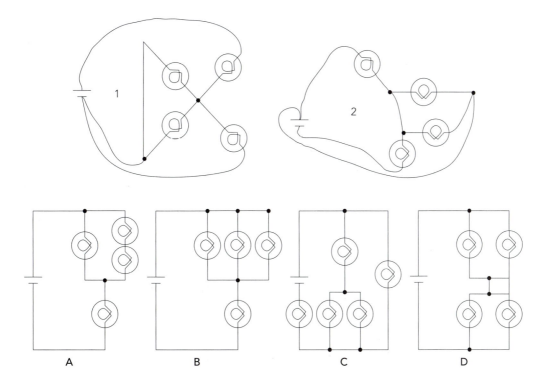

   b. Which circuit diagram (A, B, C, or D) corresponds to circuit 2? Explain the reasons for your answer.

3. Three of the circuits drawn below are electrically equivalent and one is not.

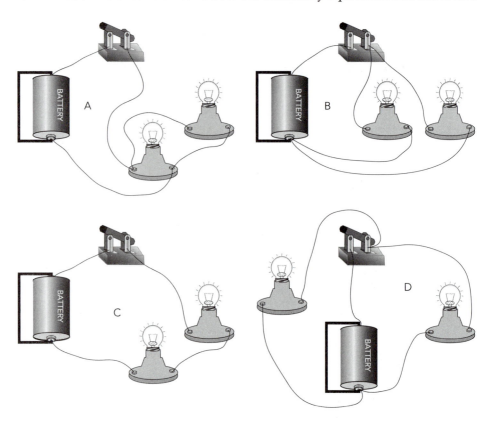

a. Which circuit is not like the others? Explain how it is different.

b. Which circuits represent parallel arrangements for the bulbs? Which represent series arrangements?

c. In the boxes below, draw neat circuit diagrams for each of the arrangements.

A	B
C	D

4. Use the model for electric current to rank the resistor networks shown below in order by resistance from largest to smallest. Explain your reasoning.

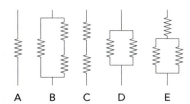

A   B   C   D   E

5. If a battery were connected to each of the circuits in Question 4, in which network would the current through the battery be the largest? The smallest? Explain your reasoning.

6. The diagram below shows a typical household circuit. The appliances (lights, television, toaster, etc.) are represented by boxes labeled 1, 2, 3, and so on. The fuse, or circuit breaker, shown in the diagram is a switch intended to shut off the circuit automatically if the wires become too hot because the current in the circuit is too large.

> **Note:** Although houses in the United States use alternating current (AC), which differs in some ways from the direct current (DC) we have been studying, you can use the model you developed for this problem. (AC circuits will be studied in Labs 7 and 8.)

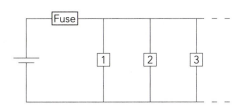

a. What happens to the current through the fuse when more appliances are added to the circuit? Describe evidence from this lab for your answer.

b. Does the current through element 1 change when elements 2 and 3 are added to the circuit? Describe evidence from this lab for your answer.

c. Is this model consistent with your observations of everyday household electricity? For example, what happens to the brightness of a light bulb in a room when a second one is turned on?

d. What may happen to the fuse if too many appliances are added to the circuit? Why?

e. What kind of circuit connection for elements 1, 2, and 3 is shown in the diagram?

7. Consider the circuit shown on the right.

   a. Are the bulbs C, D, and E connected in series, parallel or neither? Explain.

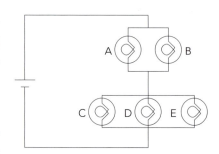

   b. Rank the bulbs in order of brightness. Use the symbols =, <, and >. Explain your ranking.

   c. How will the brightness of bulbs A and B change if bulb C is unscrewed? Will the result be different if bulb D or E is unscrewed instead? Explain.

8. Consider the circuit shown on the right. Rank the brightness of the bulbs in the circuit. Use the symbols =, <, and >. Explain your ranking.

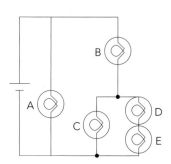

9. In the two circuits below, the batteries and all bulbs are identical. Compare the current in the circuit on the left to the current in the circuit on the right. Be as quantitative as possible.

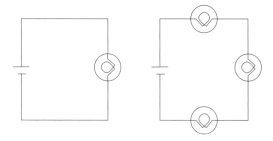

10. In the two circuits below, the batteries and all resistors are identical. Compare the current in the circuit on the left to the current in the circuit on the right. Be as quantitative as possible.

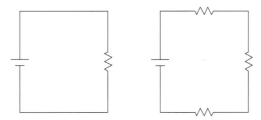

Is your answer the same as in Question 9? Explain any differences.

Name_____   Date_____   51

Pre-Lab Preparation Sheet for Lab 3—Voltage in Simple DC Circuits and Ohm's Law

(Due at beginning of lab)

Directions:
Read over Lab 3 and then answer the following questions about the procedures.

1. What do you predict for the brightness of the bulbs in Figure 3-1a and Figure 3-1c?

2. What do you predict for the brightness of the bulbs in Figure 3-2 and Figure 3-1a?

3. How will you measure the voltages between points 1 and 2 in the three circuits shown in Figure 3-3? What equipment will you use?

4. What do you predict will happen to the voltage across the battery in Figure 3-7a when you close the switch?

5. What is the function of the power supply in Figures 3-12 and 3-13?

6. How will you determine the quantitative relationship between the voltage across a resistor and the current through it? What equipment will you use?

LAB 3:
VOLTAGE IN SIMPLE DC CIRCUITS AND OHM'S LAW*

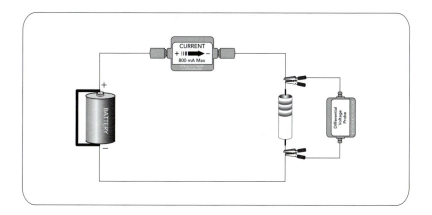

*I have a strong resistance to understanding
the relationship between voltage and current.*

—Anonymous introductory physics student

OBJECTIVES

- To learn to apply the concept of potential difference (voltage) to explain the action of a battery in a circuit.

- To understand how potential difference (voltage) is distributed in different parts of a series circuit.

- To understand how potential difference (voltage) is distributed in different parts of a parallel circuit.

- To understand the quantitative relationship between potential difference and current for a resistor (Ohm's law).

OVERVIEW

In the last two labs you explored currents at different points in series and parallel circuits. You saw that in a series circuit, *the current is the same through all elements.* You also saw that in a parallel circuit, *the current divides among the branches so that the total current through the battery equals the sum of the currents in each branch.*

You have also observed that when two or more parallel branches are connected directly across a battery, making a change in one branch does not affect the current in the other branch(es), while changing one part of a series circuit changes the current in all parts of that series circuit.

---

*Some of the activities in this lab have been adapted from those designed by the Physics Education Group at the University of Washington.

In carrying out these observations of series and parallel circuits, you have seen that connecting light bulbs in series results in a larger resistance to current and therefore a smaller current, while a parallel connection results in a smaller resistance and larger current.

In this lab, you will first examine the role of the battery in causing a current in a circuit. You will then compare the potential differences (voltages) across different parts of series and parallel circuits.

Based on your previous observations, you probably associate a larger resistance connected to a battery with a smaller current, and a smaller resistance with a larger current. In the last part of this lab you will explore the quantitative relationship between the current through a *resistor* and the potential difference (voltage) across the resistor. This relationship is known as Ohm's law.

INVESTIGATION 1: BATTERIES AND VOLTAGES IN SERIES CIRCUITS

So far you have developed a current model and the concept of resistance to explain the relative brightness of bulbs in simple circuits. Your model says that when a battery is connected to a complete circuit, there is a current. For a given battery, the magnitude of the current depends on the total resistance of the circuit. In this investigation you will explore batteries and the potential differences (voltages) between various points in circuits.

To do this you will need the following items:

- computer-based laboratory system
- two voltage probes
- *RealTime Physics Electric Circuits* experiment configuration files
- 2 1.5-V D batteries (must be very fresh, alkaline) and holders
- 6 wires with alligator clip leads
- 4 #14 bulbs in sockets
- contact switch

You have already seen what happens to the brightness of the bulb in circuit 3-1a if you add a second bulb in series as shown in circuit 3-1b. The two bulbs are not as bright as the original bulb. We concluded that the resistance of the circuit is larger, resulting in less current through the bulbs.

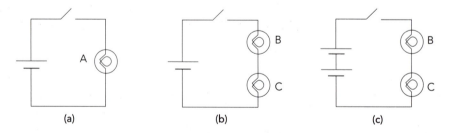

(a)   (b)   (c)

Figure 3-1: Series circuits with (a) one battery and one bulb, (b) one battery and two bulbs, and (c) two batteries and two bulbs. (All batteries and all bulbs are identical.)

Prediction 1-1: What do you predict would happen to the brightness of the bulbs if you connected a second battery in series with the first at the same time you added the second bulb as in Figure 3-1c? How would the brightness of bulb A in circuit 3-1a compare to bulb B in circuit 3-1c? To bulb C?

Activity 1-1: Battery Action

1. Connect the circuit in Figure 3-1a, and observe the brightness of the bulb.

2. Now connect the circuit in Figure 3-1c. (Be sure that the batteries are connected *in series*—the positive terminal of one must be connected to the negative terminal of the other.)

Question 1-1:   Compare the brightness of each of the bulbs to the single-bulb circuit.

Question 1-2:   What do you conclude about the current in the two-bulb, two-battery circuit as compared to the single-bulb, single-battery circuit? Explain.

Question 1-3:   What happens to the resistance of a circuit as more bulbs are added in series? What must you do to keep the current from decreasing?

Prediction 1-2:   What do you predict about the brightness of bulb D in Figure 3-2 compared to bulb A in Figure 3-1a? Explain your prediction.

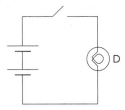

Figure 3-2: Series circuit with two batteries and one bulb.

3. Connect the circuit in Figure 3-2. *Close the switch for only a moment to observe the brightness of the bulb—otherwise, you will burn out the bulb.*

Question 1-4:   Compare the brightness of bulb D to the single-bulb circuit with only one battery (bulb A in Figure 3-1a).

Question 1-5: How does increasing the number of batteries connected in series affect the current in a series circuit?

When a battery is fresh, the voltage marked on it is actually a measure of the *emf* (*electromotive force*) or electric *potential difference* between its terminals. *Voltage* is an informal term for emf or potential difference. If you want to talk to physicists you should refer to potential difference. Communicating with a salesperson at the local Radio Shack store is another story. There you would probably refer to voltage. We will use the two terms interchangeably.

Let's explore the potential differences of batteries and bulbs in series and parallel circuits to see if we can come up with rules for them as we did earlier for currents.

How do the potential differences of batteries add when the batteries are connected in series or parallel? Figure 3-3 shows a single battery, two batteries identical to it connected in series, and two batteries identical to it connected in parallel.

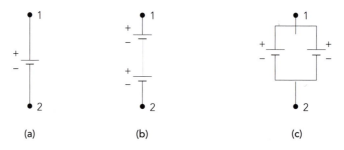

(a)   (b)   (c)

Figure 3-3: Identical batteries: (a) single, (b) two connected in series, and (c) two connected in parallel.

Prediction 1-3: If the potential difference between points 1 and 2 in Figure 3-3a is known, predict the potential difference between points 1 and 2 in Figure 3-3b (series connection) and in Figure 3-3c (parallel connection). Explain your reasoning.

Activity 1-2: Batteries in Series and Parallel

You can measure potential differences with voltage probes connected as shown in Figure 3-4.

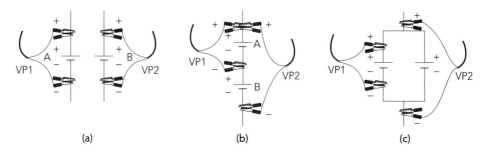

(a)   (b)   (c)

Figure 3-4: Voltage probes connected to measure the potential difference across (a) two single batteries, (b) a single battery and two batteries connected in series, and (c) a single battery and two batteries connected in parallel.

1. Open the experiment file called **Batteries (L03A1-2)**.

2. **Calibrate** the probes or **load the calibration**. **Zero** the probes with them not connected to anything.

3. Connect voltage probe 1 across a single battery (as in Figure 3-4a), and voltage probe 2 across the other identical battery.

4. Record the voltage measured for each battery below.

    Voltage of battery A:_____   Voltage of battery B:_____

Question 1-6: How do your measured values agree with those marked on the batteries?

5. Now connect the batteries in series as in Figure 3-4b, and connect probe 1 to measure the potential difference across battery A and probe 2 to measure the potential difference across the series combination of the two batteries. Record your measured values below.

    Voltage of battery A:_____   Voltage of A and B in series:_____

Question 1-7: Do your measured values agree with your predictions? Can you explain any differences?

6. Now connect the batteries in parallel as in Figure 3-4c, and connect probe 1 to measure the potential difference across battery A and probe 2 to measure the potential difference across the parallel combination of the two batteries. Record your measured values below.

    Voltage of battery A:_____   Voltage of A and B in parallel:_____

Question 1-8: Do your measured values agree with your predictions? Can you explain any differences?

Question 1-9: Make up a rule for finding the combined voltage of a number of batteries connected in series.

Question 1-10: Make up a rule for finding the combined voltage of a number of identical batteries connected in parallel.

You can now explore the potential difference across different parts of a simple series circuit. Let's begin with the circuit with two bulbs in series with a battery, which you looked at before in Lab 2, Activities 1-1 and 1-2. It is shown in Figure 3-5a.

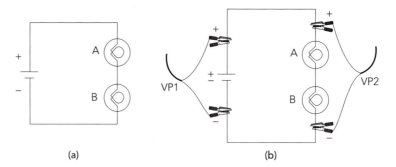

Figure 3-5: (a) A series circuit with one battery and two bulbs, and (b) the same circuit with voltage probe 1 connected to measure the potential difference across the battery and probe 2 connected to measure the potential difference across the series combination of bulbs A and B.

Prediction 1-4: If bulbs A and B are identical, predict how the potential difference (voltage) across bulb A in Figure 3-5b will compare to the potential difference across the battery. How about bulb B? How about the potential difference across the series combination of bulbs A and B—how will this compare to the voltage across the battery?

Test your prediction.

Activity 1-3: Voltages in Series Circuits

1. Open the experiment file called **Batteries (L03A1-2)**, if it is not already open.

2. **Calibrate** the voltage probes or **load the calibration**, if this has not already been done. **Zero** both probes with nothing connected to them.

3. Connect the circuit shown in Figure 3-5b.

> **Comment:** In carrying out your measurements, remember that the measurements made by the voltage probes are only as good as their calibration. Small differences in calibration can result in small differences in readings.

4. Measure the voltages, and record your readings below.

   Potential difference across the battery:_____

   Potential difference across bulbs A and B in series:_____

Question 1-11: How do the two potential differences compare? Did your observations agree with your predictions?

5. Connect the voltage probes as in Figure 3-6 to measure the potential difference across bulb A and across bulb B. Record your measurements below.

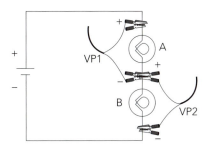

Figure 3-6: Connection of voltage probes to measure the potential difference across bulb A and across bulb B.

Potential difference across bulb A:_____

Potential difference across bulb B:_____

Question 1-12: Did your measurements agree with your predictions?

Question 1-13: Formulate a rule for how potential differences across individual bulbs in a series connection combine to give the total potential difference across the series combination of the bulbs. How is this related to the potential difference of the battery?

INVESTIGATION 2: VOLTAGES IN PARALLEL CIRCUITS

You can also explore the potential differences across different parts of a simple *parallel* circuit. Let's begin with the circuit with two bulbs in parallel with a battery, which you looked at in Lab 2. It is shown in Figure 3-7a.

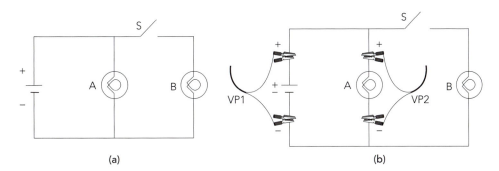

(a)   (b)

Figure 3-7: (a) Parallel circuit with two bulbs and a battery, and (b) the same circuit with voltage probe 1 connected to measure the potential difference across the battery and probe 2 connected to measure the potential difference across bulb A.

Prediction 2-1: What do you predict will happen to the potential difference across the battery when you close the switch in Figure 3-7a? Will it increase, decrease, or remain essentially the same? Explain.

Prediction 2-2: With the switch in Figure 3-7a closed, how will the potential difference across bulb A compare to the voltage of the battery? How will the potential difference across bulb B compare to the voltage of the battery?

To test your predictions you will need:

- computer-based laboratory system
- 2 voltage probes
- *RealTime Physics Electric Circuits* experiment configuration files
- 1.5-V D battery (must be very fresh, alkaline) with holder
- 6 alligator clip leads
- 2 #14 bulbs in sockets
- contact switch

Activity 2-1: Voltages in a Parallel Circuit

1. The experiment file **Batteries (L03A1-2)** should still be open, and the axes that follow should be on the screen.

2. **Calibrate** the voltage probes or **load the calibration**, if this has not already been done. **Zero** both probes with nothing connected to them.

3. Connect the circuit shown in Figure 3-7b.

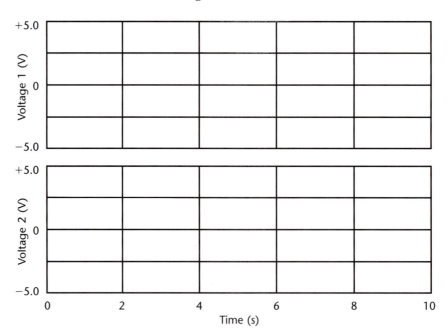

4. **Begin graphing,** and then close and open the switch as you've done before.

5. Sketch the graphs on the axes above, or **print** and affix them over the axes.

6. Read the voltages using the **analysis feature** of the software.

    Switch open: Voltage across battery:_____ Voltage across bulb A:_____

    Switch closed: Voltage across battery:_____ Voltage across bulb A:_____

Question 2-1: Did your measurements agree with your predictions? Did closing and opening the switch significantly affect the voltage across the battery (by more than several percent)? The voltage across bulb A?

7. Now connect the voltage probes as shown in Figure 3-8, and graph and measure the voltages across bulbs A and B. Again close and open the switch while graphing.

8. Sketch your graphs or **print** them and affix them over the axes.

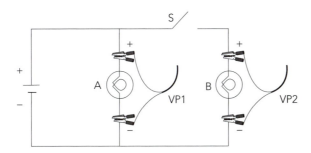

Figure 3-8: Voltage probes connected to measure the potential differences across bulbs A and B.

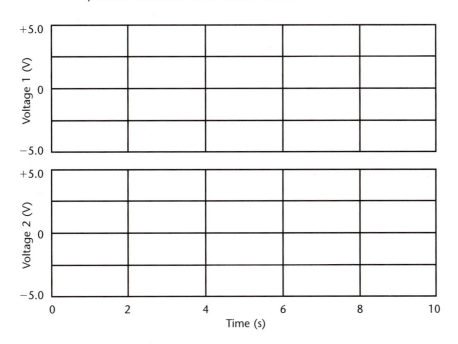

9. Record your measurements using the **analysis feature** of the software.

Switch open: Voltage across bulb A:_____ Voltage across bulb B:_____

Switch closed: Voltage across bulb A:_____ Voltage across bulb B:_____

Question 2-2: Did your measurements agree with your predictions? Did closing and opening the switch significantly affect the voltage across bulb A (by more than several percent)?

Question 2-3: Did closing and opening the switch significantly affect the voltage across bulb B (by more than several percent)? Under what circumstances is there a potential difference across a bulb?

Question 2-4: Based on your observations, formulate a rule for the potential differences across the different branches of a parallel circuit. How are these related to the voltage across the battery?

Question 2-5: Is a battery a constant current source (delivering essentially a fixed amount of current regardless of the circuit connected to it) or a constant voltage source (applying essentially a fixed potential difference regardless of the circuit connected to it), or neither? Explain based on your observations in this and the previous lab.

Question 2-6: What is the voltage between two points on a short length of wire when there is no bulb, battery, or resistor between the points?

You have now observed several times in these activities that the voltage across a very fresh alkaline battery doesn't change significantly no matter what is connected to the battery (no matter how much current flows in the circuit). As you will see in the following activity, this is not true for a less than fresh battery.

In addition to the materials you have been using, you will need

- 1.5-V D battery that is not very fresh
- additional #14 bulbs in socket
- additional contact switch

Activity 2-2: Internal Resistance of a Battery

1. Open the experiment file **Internal Resistance (L03A2-2)** to measure voltage with voltage probe 1 and current with current probe 2.

2. **Calibrate** the probes or **load the calibration**, if this has not already been done. **Zero** both probes with nothing connected to them.

3. Connect the circuit shown in Figure 3-9.

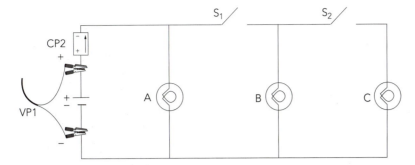

Figure 3-9: Circuit to examine voltage across a not-so-fresh battery as the current through the battery increases.

4. Measure the voltage across the battery and the current through the battery with both switches open, with S1 closed and with both switches closed (Table 3-1).

Table 3-1

	Voltage (V)	Current (A)
Both switches open		
S1 closed		
Both switches closed		

Question 2-7: Did the voltage across this not-so-fresh battery remain constant as the current through the battery increased? If not, how did it change?

Batteries are sources of potential energy for the charges flowing through them. They also have an *internal resistance* that increases in size as they wear out. The equivalent circuit of a battery with internal resistance is shown in Figure 3-10.

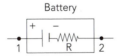

Figure 3-10: Equivalent circuit of a battery with internal resistance.

Question 2-8: Are your measurements for voltage and current in Table 3-1 consistent with the equivalent circuit in Figure 3-10? Explain.

If you have time, work on the following extension.

Extension 2-3: Applying Your Current and Voltage Models

Let's return to a more complex circuit using what we now know about voltage and current. In Lab 2, Investigation 3, you explored the circuit shown in Figure 3-11.

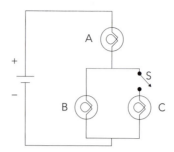

Figure 3-11: Circuit equivalent to Figure 2-9a when the switch is open, and to Figure 2-9b when the switch is closed.

You were previously asked to rank the brightness of bulbs A, B, and C after the switch was closed. The question now is, *what happens to the brightness of bulb B when the switch is closed?* Does it increase, decrease or remain the same?

Prediction E2-3: Based on the current and voltage models you have developed, *carefully* predict what will happen to the current through bulb B (and therefore its brightness) when bulb C is added in parallel to it. Will it increase, decrease, or remain the same? Explain the reasons for your answer.

Connect the circuit in Figure 3-11, and make observations. Describe what happens to the brightness of bulb B when the switch is closed.

Question E2-9: Did your observations agree with your prediction? If not, use the current and voltage models to explain your observations.

INVESTIGATION 3: OHM'S LAW

What is the relationship between current and potential difference? You have already seen on several occasions that there is only a potential difference across a bulb or resistor when there is a current through the circuit element. The next question is how does the potential difference depend on the current? To explore this, you will need the following:

- computer-based laboratory system
- current and voltage probes
- *RealTime Physics Electric Circuits* experiment configuration files
- variable regulated DC power supply (up to 3 V and 0.5 amps)
- 6 alligator clip leads
- 10-Ω resistor
- #14 bulb in a socket

Examine the circuit shown in Figure 3-12. A variable DC power supply is like a variable battery. When you turn the dial, you change the voltage (potential difference) between its terminals. Therefore, this circuit allows you to measure the current through the resistor when different voltages are across it.

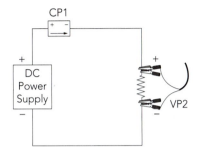

Figure 3-12: Circuit with a variable power supply to explore the relationship between current and potential difference for a resistor.

Prediction 3-1: What will happen to the current *through the resistor* as you turn the dial on the power supply and increase the applied voltage from zero?

Prediction 3-2: What will happen to the potential difference *across the resistor* as the current through it increases from zero?

Prediction 3-3: What will be the mathematical relationship between the *voltage across the resistor* and the *current through the resistor*?

Activity 3-1: Current and Potential Difference for a Resistor

1. Open the experiment file called **Ohm's Law (L03A3-1).**

2. **Calibrate** the current and voltage probes or **load the calibration**, if this has not already been done. **Zero** both probes with nothing connected to them.

3. Connect the circuit in Figure 3-12. Note that the current probe is connected to measure the current through the resistor, and the voltage probe is connected to measure the potential difference across the resistor.

   Your instructor will show you how to operate the power supply.

4. **Begin graphing** current and voltage with the power supply set to zero voltage, and graph as you turn the dial and increase the voltage *slowly* to about 3 V.

Warning: Do not exceed 3 V!

Question 3-1: What happened to the current in the circuit as the power supply voltage was increased? Did this agree with your prediction?

Question 3-2: How did the potential difference across the resistor change as the current through the resistor changed? Did this agree with your prediction?

5. You can display axes for voltage vs. *current* on the bottom graph by **adjusting the horizontal axis** to read **Current 1**. The axes should now be as shown below.

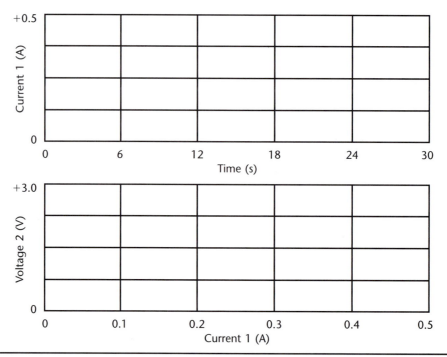

Reminder: We are interested in the nature of the mathematical relationship between the voltage across the resistor and the current through the resistor. This can be determined from the graph by drawing a smooth curve that fits the plotted data points. Some definitions of possible mathematical relationships are shown below. In these examples, *y* might be the voltage reading and *x* the current.

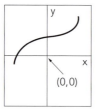

y is a function of x which increases as x increases.

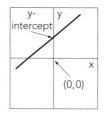

y is a *linear* function of x which increases as x increases according to the mathematical relationship y = mx + b, where b is a constant called the *y*-intercept.

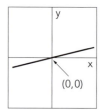

y is *proportional* ti x. This is a special case of a linear relationship where y = mx, and b, the *y-intercept*, is zero.

These graphs show the differences between these three types of mathematical relationship. *y* can increase as *x* increases, and the relationship doesn't have to be *linear* or *proportional*.

Proportionality refers only to the special linear relationship where the *y*-intercept is zero, as shown in the example graph on the right.

6. Use the **fit routine** in the software to see if the relationship between voltage and current for a resistor is a proportional one.

7. Sketch your graphs on the axes above, or **print** them and affix them over the axes.

Question 3-3: In words, what is the mathematical relationship between potential difference and current for a resistor? Explain based on your graphs.

The relationship between potential difference and current that you have observed for a resistor is known as Ohm's law. To put this law in its normal form, we must now define the quantity known as *resistance*. Resistance is defined as the slope of the voltage vs. current graph.

If potential difference is measured in volts and current is measured in amperes, then the unit of resistance is the ohm, which is usually represented by the Greek letter "omega" (Ω).

Question 3-4: State the mathematical relationship that you determined from the fit to your graph, in terms of V, I, and R.

Question 3-5: Based on your graph, what can you say about the value of R for a resistor—is it constant or does it change as the current through the resistor changes? Explain.

Question 3-6: From the slope of your graph, what is the experimentally determined value of the resistance of your resistor in ohms? How does this agree with the value written on the resistor? (Remember the tolerance.)

> **Note:** Many circuit elements do not obey Ohm's law. The definition for resistance is still the same, but the resistance changes as the current changes. Circuit elements that follow Ohm's law—like resistors—are said to be *ohmic*.

In the last activity you explored the relationship between the potential difference across a resistor and the current through the resistor. It is a proportional relationship. Instead of a resistor, in the following extension you will explore the relationship between current and potential difference for a light bulb.

Extension 3-2: Relationship Between Current and Potential Difference for a Light Bulb

1. Replace the 10-Ω resistor by the light bulb (Figure 3-13).

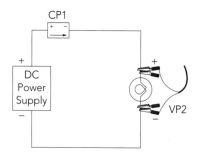

Figure 3-13: Circuit with a variable power supply to explore the quantitative relationship between the current and potential difference for a light bulb.

Prediction E3-4: What do you predict will happen to the brightness of the bulb as you turn the dial on the power supply and increase the voltage from zero? Explain.

Prediction E3-5: What will be the mathematical relationship between the *voltage across the bulb* and the *current through the bulb*?

2. Prepare to graph current vs. time with probe 1 and voltage vs. time with probe 2. (**Adjust the horizontal axis** on the bottom graph back to **time**.)

3. **Begin graphing** with the power supply set to zero voltage, and graph current and voltage as you turn the dial and increase the voltage *slowly* to about 3 V.

Warning: Do not exceed 3 V since this may burn out the bulb!

Question E3-7: What happened to the brightness of the bulb as the power supply voltage was increased? Did this agree with your prediction?

Question E3-8: How is the brightness of the bulb related to the potential difference across the bulb? To the current through the bulb?

4. You can again display a graph of potential difference vs. current by **adjusting the horizontal axis** on the bottom graph to **Current 1** as before.

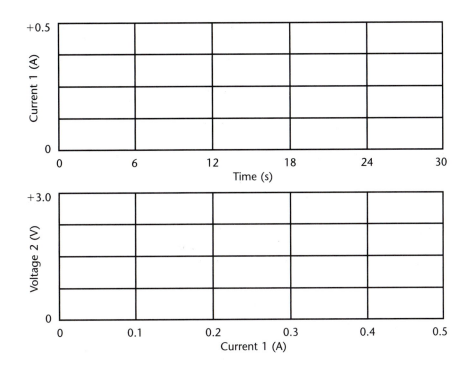

Question E3-9: Compare your graph of voltage vs. current for the bulb to that for the resistor in the previous activity. In what ways are they similar and in what ways are they different?

5. Use the **fit routine** in the software to determine if the relationship between voltage and current for a light bulb is a proportional one.

6. Sketch the graphs, or **print** and affix them over the axes.

Question E3-10: Based on your graph of voltage vs. current for a bulb, what can you say about the value of R for a bulb—is it constant or does it change as the current through the bulb changes? Explain.

Question E3-11: Is a light bulb an ohmic device? Explain.

HOMEWORK FOR LAB 3
VOLTAGE IN SIMPLE DC CIRCUITS AND OHM'S LAW

1. In the circuit below, the battery maintains a constant potential difference between its terminals at points 1 and 2 (i.e., the internal resistance of the battery is considered negligible).

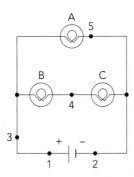

The three light bulbs, A, B, and C are identical.

   a. How do the brightnesses of the three bulbs compare to each other? Explain your reasoning.

   b. What happens to the brightness of each of the three bulbs when bulb A is unscrewed and removed from its socket? Explain your reasoning.

   c. When A is unscrewed, what happens to the current through points 3, 4, and 5? Explain your reasoning.

   d. Bulb A is screwed back in. What happens to the brightness of each of the three bulbs when bulb C is unscrewed and removed from its socket? Explain your reasoning.

   e. When C is unscrewed, what happens to the current through points 3, 4, and 5? Explain your reasoning.

2. For each of the questions A–E below, a wire is connected from the battery terminal at point 1 to point 4.

   a. What happens to the brightness of each of the three bulbs? Explain.

b. What happens to the current through point 3? Explain.

c. What happens to the potential difference across bulb B? Explain.

d. What happens to the potential difference across bulb C? Explain.

e. What happens to the potential difference between points 1 and 5? Explain your reasoning.

3. The wire in (2) is removed. What happens to the brightness of each of the three bulbs and to the current through point 2 if a wire is connected from the battery terminal at point 2 to the socket terminal at point 5?

4. The circuit is returned to its original state. A fourth bulb (D) is connected in parallel with bulb B (*not in parallel with B and C*).

   a. Sketch the bulb in the circuit.

   b. What happens to the brightness of each of the three bulbs?

   c. What happens to the current through point 3?

   d. What happens to the potential difference between points 3 and 4?

   e. What happens to the potential difference between points 4 and 2?

5. State Ohm's law in words. For what type of circuit elements does it correctly describe the behavior?

6. Does a light bulb have a constant resistance? Explain. Is a light bulb *ohmic*?

7. Does a resistor have a constant resistance? Explain. Is a resistor *ohmic*?

8. Draw diagrams for a 75-Ω and a 100-Ω resistor connected in series and connected in parallel:

        Series   Parallel

9. In the following circuits, tell which resistors are connected in series, which are connected in parallel, and which are neither in series nor parallel.

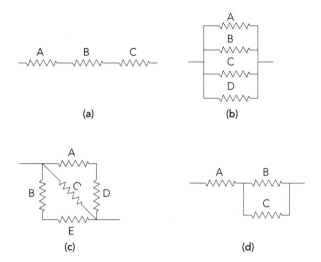

(a)   (b)

(c)   (d)

PRE-LAB PREPARATION SHEET FOR LAB 4—KIRCHHOFF'S CIRCUIT RULES

(Due at the beginning of lab)

Directions:
Read over Lab 4 and then answer the following questions about the procedures.

1. What is a multimeter? What can quantities you measure with one?

2. How should a multimeter be connected to measure the potential difference across a resistor—in series or in parallel with the resistor?

3. A resistor has four colored stripes in the following order: orange, violet, red, and silver. What is the resistance and the tolerance?

4. How will you measure resistances to determine a rule for adding resistors in parallel?

5. State Kirchhoff's Loop Rule.

6. State Kirchhoff's Junction Rule.

7. Which of Kirchhoff's rules is based on conservation of charge?

LAB 4:
KIRCHHOFF'S CIRCUIT RULES

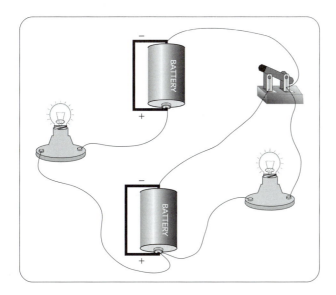

The Energizer© keeps on going, and going, and . . . .

—Eveready Battery Company, Inc.

OBJECTIVES

- To learn how multimeters are designed to minimize interference with the currents and voltages to be measured.

- To learn how to measure resistance with a multimeter.

- To discover the rule for determining the equivalent resistance of resistors connected in series.

- To discover the rule for determining the equivalent resistance of resistors connected in parallel.

- To examine Kirchhoff's circuit rules and apply them to some DC circuits.

OVERVIEW

In the last few labs, you have examined simple circuits involving series and parallel connections of light bulbs and resistors. The emphasis has been on understanding the concepts of current, voltage, and resistance in DC circuits.

In this lab you will look at circuits more quantitatively. Up until now you have been using computer-based probes to measure currents and voltages. A *multimeter* is a device with the capability of measuring current and voltage. Some multimeters can also be used to measure resistance. In Investigation 1, you will learn how to measure current, voltage, and resistance with a multimeter.

In Investigation 2, you will look at some circuits consisting of resistors connected in series or in parallel and develop the rules for combining resistances in series and parallel.

Often circuit elements are connected with multiple batteries in more complicated ways than simply in series or in parallel. The rules for series and parallel addition of resistances often are not adequate to examine what is going on in such circuits. In Investigation 3 of this lab, you will examine Kirchhoff's circuit rules that are generally applicable to all types of circuits.

INVESTIGATION 1: MEASURING CURRENT, VOLTAGE, AND RESISTANCE

The multimeters available to you can be used to measure current, voltage, or resistance. All you need to do is choose the correct dial setting, connect the wire leads to the correct terminals on the meter, and connect the meter correctly in the circuit. Figure 4-1 shows a simplified diagram of a multimeter.

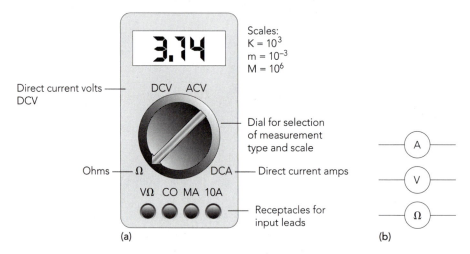

Figure 4-1: (a) Multimeter with voltage, current, and resistance modes, and (b) symbols that will be used to indicate a multimeter used as an ammeter, voltmeter, and ohmmeter respectively.

A current probe and a multimeter used to measure current are both connected in a circuit in the same way. Likewise, a voltage probe and a multimeter used to measure voltage are both connected in a circuit in the same way. The next two activities will remind you how to connect them. The activities will also show you that when meters are connected correctly, they don't interfere with the currents or voltages being measured.

You will need:

- digital multimeter
- 2 1.5-V D batteries (must be very fresh, alkaline) with holders
- 6 alligator clip leads
- 2 #14 bulbs and sockets

Activity 1-1: Measuring Current with a Multimeter

Figure 4-2 shows two possible ways that you might connect a multimeter to measure the current through bulb 1.

Prediction 1-1: Which of the diagrams in Figure 4-2, (b) or (c), shows the correct way to connect a multimeter to measure the current through bulb 1? Explain why it should be connected this way. [**Hint:** In which case is the current through the multimeter the same as that through bulb 1?]

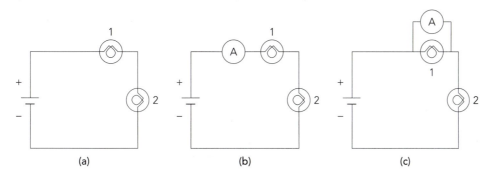

Figure 4-2: A circuit with two light bulbs and a battery (a), and two possible *but not necessarily desirable* ways to connect a multimeter to measure current: (b) in series with bulb 1, and (c) in parallel with bulb 1.

1. Set up the basic circuit in Figure 4-2a. Use the two batteries in series to make a 3-V battery. Observe the brightness of the bulbs.

2. Set the multimeter to measure current and connect it as shown in Figure 4-2b. Was the brightness of the bulbs significantly affected?

3. Now connect the meter as in Figure 4-2c. Was the brightness of the bulbs significantly affected?

Question 1-1: If the multimeter is connected correctly to measure current, it should measure the current through bulb 1 without significantly affecting the current through the bulb. Which circuit in Figure 4-2 shows the correct way to connect a multimeter? Explain based on your observations. Why is it connected in this way?

Question 1-2: Does the multimeter appear to behave as if it is a large or small resistor? Explain based on your observations. Why is it designed in this way?

Activity 1-2: Measuring Voltage with a Multimeter

Figure 4-3 shows two possible ways that you might connect a multimeter to measure the potential difference across bulb 1.

Prediction 1-2: Which of the diagrams in Figure 4-3 shows the correct way to connect a multimeter to measure the voltage across bulb 1? Explain why it should be connected this way.

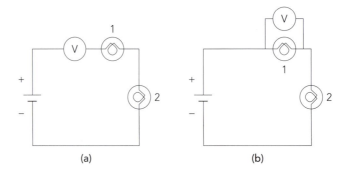

Figure 4-3: Two possible *but not necessarily desirable* ways to connect a multimeter to measure voltage: (a) in series with bulb 1, and (b) in parallel with bulb 1.

1. Set the meter to measure voltage and connect it as in Figure 4-3a. Was the brightness of the bulbs significantly affected?

2. Now connect the meter as in Figure 4-3b. Was the brightness significantly affected?

Question 1-3: If the multimeter is connected correctly, it should measure the voltage across bulb 1 without significantly affecting the current through the bulb. Which circuit in Figure 4-3 shows the correct way to connect the multimeter? Explain based on your observations. Why is it connected in this way? [**Hint:** In which case is the voltage across the multimeter the same as that across bulb 1?]

Question 1-4: Does the multimeter behave as if it is a large or small resistor? Explain based on your observations. Why is it designed in this way?

Next we will investigate how to measure resistance with a multimeter. Resistance, voltage, and current are fundamental electrical quantities that characterize all electric circuits. You just observed that even multimeters exhibit some amount of resistance.

In earlier labs, you observed that light bulbs exhibit resistance that increases with the current through the bulb (with the temperature of the filament). To make the design and analysis of circuits as simple as possible, it is desirable to have circuit elements with resistances that do not change. For that reason, *resistors* are used in electric circuits. The resistance of well-designed resistors doesn't vary with the amount of current passing through them (or with the temperature), and they are inexpensive to manufacture.

One type of resistor is a carbon resistor. It contains a form of carbon known as graphite suspended in a hard glue binder. It is usually surrounded by a plastic case with a color code painted on it. Look at the samples of carbon resistors that have been cut down the middle as shown in Figure 4-4.

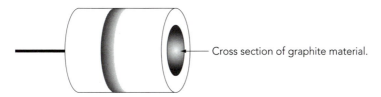

Cross section of graphite material.

Figure 4-4: A cutaway view of a carbon resistor.

Figure 4-5 shows a carbon resistor with colored bands that tell you the value of the resistance and the tolerance (guaranteed accuracy) of this value.

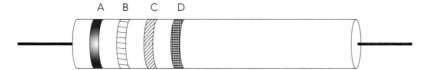

A   B   C   D

Figure 4-5: A carbon resistor with color bands.

The first two stripes indicate the two digits of the resistance value. The third stripe indicates the power of ten multiplier, and the fourth stripe tells the tolerance. The key in Table 4-1 shows the corresponding values.

Table 4-1: The resistor code	
Bands 1–3	violet = 7
black = 0	gray = 8
brown = 1	white = 9
red = 2	
orange = 3	*Band 4*
yellow = 4	none = ±20%
green = 5	silver = ±10%
blue = 6	gold = ±5%

As an example, look at the resistor in Figure 4-6. Its two digits are 1 and 2 and the multiplier is 10^3, so its value is 12×10^3, or 12,000 Ω. The tolerance is ±20%, so the value might actually be as large as 14,400 Ω or as small as 9600 Ω.

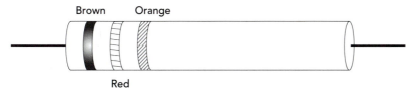

Brown   Orange

Red

Figure 4-6: An example of a color-coded carbon resistor. The resistance of this resistor is 12×10^3 Ω ± 20%.

The connection of the multimeter to measure resistance is shown in Figure 4-7. When the multimeter is in its ohmmeter mode, it connects a known voltage across the resistor and measures the current through the resistor. Then resistance is calculated by the meter from $R = V/I$.

Note: Resistors must be isolated (disconnected from the circuit) before their resistances can be measured. This also prevents damage to the multimeter that may occur if a voltage is connected across its terminals while in the resistance mode.

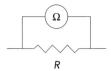

Figure 4-7: Connection of an ohm-meter to measure resistance.

In the next activity, you will use the multimeter to measure the resistance of several resistors. You will need

- several color-coded resistors
- 2 digital multimeters
- 6-V battery

Activity 1-3: Reading Resistor Codes and Measuring Resistance

1. Choose several resistors and read their codes. Record the resistances and the tolerances in the first two columns of Table 4-2.

Table 4-2

R from code (Ω)	Tolerance from code	Measured R (Ω)	Measured V (V)	Measured I (A)	$R = V/I$ (Ω)

There are two ways to determine resistance with the multimeter. One is to use the resistance mode and measure the resistance directly. The second is to connect the resistor to a battery and then measure the voltage across the resistor and the current through the resistor. R can then be calculated from Ohm's law, $R = V/I$. You will use both of these methods to measure the resistances of the resistors you selected.

2. Measure the resistance of each of your resistors directly using the multimeter and enter the values in Table 4-2 under "Measured R."

3. *Be sure to reset the multimeters to measure current and voltage respectively, or you will burn out the fuse in the meter.* Connect each resistor to the battery and simultaneously measure the voltage across the resistor and the current through the resistor with the multimeters. Record V and I in the appropriate columns Table 4-2 and calculate the resistance from these values.

Question 1-5: How do the values of your resistors measured with the resistance mode of the multimeter compare to the values indicated by the code? Assuming that your measured values are correct, are the values indicated by the code correct within the stated tolerance?

Question 1-6: How do the resistance values found from the voltage across the resistor and the current through it compare to the values measured with the resistance mode of the multimeter? Do you conclude that the resistance mode is reliable?

INVESTIGATION 2: SERIES AND PARALLEL COMBINATIONS OF RESISTORS

Several resistors can be wired in *series* to increase their effective length and in *parallel* to increase their effective cross-sectional area as shown in Figure 4-8.

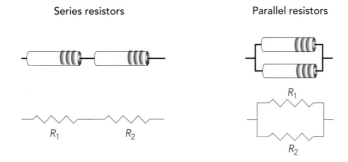

Figure 4-8: Resistors wired in series and in parallel.

To do some exploration of equivalent resistance of different resistors wired in combination you will need the following:

- 3 51-Ω resistors

- a 22-Ω and a 75-Ω resistor

- digital multimeter

- 6 alligator clip leads

Activity 2-1: Equivalent Resistance of Resistors Connected in Series

1. Measure the actual values of the 51-Ω resistors with the multimeter. Record their values below:

    R_1 _____ Ω R_2 _____ Ω R_3 _____ Ω

2. Connect R_1 and R_2 in series. Measure the resistance of this series combination with the multimeter.

    Resistance of R_1 and R_2 in series:_____ Ω

3. Now connect R_1, R_2, and R_3 in series and measure the resistance of the combination.

    Resistance of R_1, R_2, and R_3 in series:_____ Ω

Question 2-1: Based on your measurements, state a rule for finding the equivalent resistance of several resistors connected in series. If the resistors have resistances R_1, R_2, and R_3, write a mathematical equation for the equivalent resistance, R_{eq}, when these are connected in series. Explain how your measurements support this rule.

Question 2-2: Does your rule agree with your observations in Lab 2 that the current through two identical resistors connected in series is half the current through a single resistor connected to the same battery? Explain.

Activity 2-2: Equivalent Resistance of Resistors Connected in Parallel

1. Use your three 51-Ω resistors again. Connect R_1 and R_2 in parallel. Measure the resistance of this parallel combination with the multimeter.

    Resistance of R_1 and R_2 in parallel:_____ Ω

2. Now connect R_1, R_2, and R_3 in parallel, and measure the resistance of the combination.

    Resistance of R_1, R_2, and R_3 in parallel:_____ Ω

Question 2-3: Based on your measurements, is the equivalent resistance, R_{eq}, consistent with the following mathematical relationship?

$$\frac{1}{R_{eq}} = \frac{1}{R_1} + \frac{1}{R_2} + \frac{1}{R_3}$$

Show all calculations you use to check this.

Question 2-4: Does the relationship in Question 2-3 agree with your observations in Lab 2 that the current through a battery connected to two identical resistors connected in parallel is twice the current through the battery when connected to a single resistor? Explain.

Extension 2-3: Other Combinations of Resistors in Series and Parallel

1. Measure the actual values of the 22-Ω and 75-Ω resistors with the multimeter. Record their values below:

   R_4 _____ Ω R_5 _____ Ω

2. Connect R_1, R_4, and R_5 in series. Measure the resistance of this series combination.

   Resistance of R_1, R_4, and R_5 in series:_____ Ω

Question E2-5: Use your rule in Question 2-1 to calculate the equivalent resistance of these three resistors. Show your calculation. How do these values compare to the measured resistance of the series combination?

3. Connect R_1, R_4, and R_5 in parallel. Measure the resistance of this parallel combination.

   Resistance of the combination:_____ Ω

Question E2-6: Use the rule in Question 2-3 to calculate the equivalent resistance of these three resistors. Show your calculation. How do these values compare to the measured resistance of the parallel combination?

Now that you know the basic rules to calculate equivalent resistance for series and parallel connections of resistors, you can tackle the question of how to find the equivalent resistance for complex networks of resistors. The trick is to be able to calculate the equivalent resistance of each segment of the complex network and use that in calculations of the next segment.

For example, in the network shown in Figure 4-9, there are two resistance values, R_1 and R_2. A series of simplifications is shown in the diagram below.

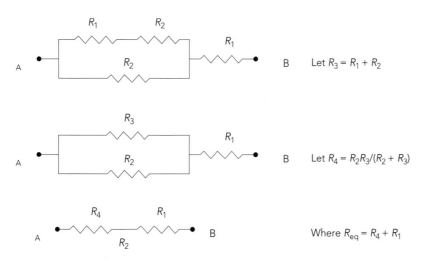

Figure 4-9: A sample resistor network.

To complete the following equivalent resistance extension you will need the following:

- 3 22-Ω resistors

- 3 51-Ω resistors

- digital multimeter

Extension 2-4: The Equivalent Resistance for a Network

1. Use the color-coded value for your smaller resistors for R_1 and the color-coded value for your larger resistors for R_2. List these values below.

    R_1:_____ R_2:_____

2. Calculate the equivalent resistance between points A and B for the network shown below. Show your calculations on a step-by-step basis.

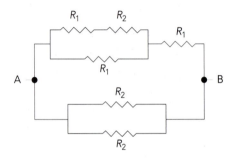

3. Set up this network of resistors and check your calculation by measuring the equivalent resistance directly with the multimeter.

Question E2-7: How did your measured value for the equivalent resistance agree with the calculated value? Could any disagreement be explained by the tolerances in the resistor values? Explain in detail.

INVESTIGATION 3: KIRCHHOFF'S CIRCUIT RULES

Suppose you want to calculate the currents in various branches of a circuit that has many components wired together in a complex array. The rules for combining resistors that you examined in Investigation 2 are very convenient in circuits made up only of resistors that are connected in series or in parallel. But, while it may be possible in some cases to simplify parts of a circuit with the series and parallel rules, complete simplification to an equivalent resistance is often impossible, especially when components other than resistors are included. The application of Kirchhoff's circuit rules can help you to understand the most complex circuits.

Before summarizing these rules, we need to define the terms *junction* and *branch*. Figure 4-10 illustrates the definitions of these two terms for an arbitrary circuit. As shown in Figure 4-10a, a junction in a circuit is a place where two or more wires are connected together. As shown in Figure 4-10b, a branch is a portion of the circuit in which the current is the same through every circuit element. (That is, the circuit elements in a branch are all connected in series with each other.)

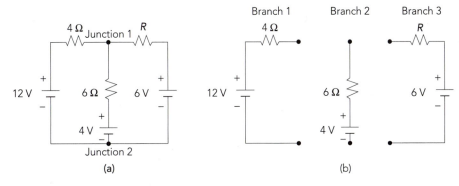

Figure 4-10: An arbitrary circuit used to illustrate junctions and branches.

Kirchhoff's rules can be summarized as follows:

---

1. *Junction Rule (based on charge conservation):* The sum of all the currents entering any junction of the circuit must equal the sum of the currents leaving.

2. *Loop Rule (based on energy conservation):* Around any closed loop in a circuit, the sum of all changes in potential (emfs and potential drops across resistors and other circuit elements) must equal zero.

These rules are applied to a circuit in the following steps:

---

1. Assign a current symbol to each branch of the circuit, and label the current in each branch I_1, I_2, I_3, etc.

2. *Arbitrarily* assign a direction to each current. (The direction chosen for the current in each branch is arbitrary. If you choose the right direction, when you solve the equations, the current will come out positive. If you choose the wrong direction, the current will come out negative, indicating that its direction is actually opposite to the one you chose.) Remember that the current is always the same everywhere in a branch, and the current out of a battery is always the same as the current into a battery.

3. Apply the *Loop Rule* to each of the loops.

   a. Let the potential drop (voltage) across each resistor be the negative of the product of the resistance and the net current through the resistor. (However, make the sign positive if you are traversing a resistor in the direction opposite that of the current).

   b. Assign a positive potential difference when the loop traverses from the − to the + terminal of a battery. (If you are going across a battery in the opposite direction assign a negative potential difference to the trip across the battery terminals.)

4. Find each of the junctions and apply the *Junction Rule* to it. You can write currents leaving the junction on one side of the equation and currents coming into the junction on the other side of the equation.

To illustrate the application of the rules let's consider the circuit in Figure 4-11.

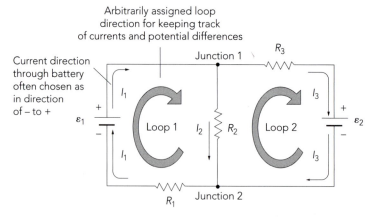

Figure 4-11: A complex circuit in which loops 1 and 2 share the resistor R_2.

Question 3-1: Why are the resistors R_1 and R_2 not in series? Why are they not in parallel?

In Figure 4-11 the directions for the loops through the circuits and for the three currents are assigned arbitrarily. If we assume that the internal resistances of the batteries are negligible, then by applying the *Loop Rule* we find that

Loop 1	$\varepsilon_1 - I_2R_2 - I_1R_1 = 0$	(1)
Loop 2	$-\varepsilon_2 + I_2R_2 - I_3R_3 = 0$	(2)

By applying the *Junction Rule* to junction 1 or 2, we find that

$$I_1 = I_2 + I_3 \qquad (3)$$

(current into junction = current out of junction)

It may trouble you that the current directions and directions that the loops are traversed have been chosen arbitrarily. You can explore this assertion by changing these choices and analyzing the circuit again. To do the following activity you'll need a couple of resistors and a multimeter as follows:

- 2 resistors (rated values of 39 and 75 Ω)
- digital multimeter (to measure resistance)
- 6-V battery
- 1.5-V D battery (very fresh, alkaline) and holder

Activity 3-1: Applying the Loop and Junction Rules Several Times

Figure 4-12 shows the same circuit as in Figure 4-11 with different arbitrary directions for the loops and current through R_2.

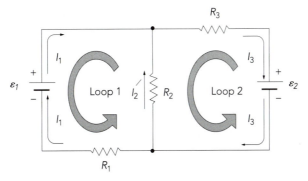

Figure 4-12: The same complex circuit as in Figure 4-11 with the current through R_2 chosen arbitrarily in the opposite direction, and the loops traversed in the counterclockwise direction instead of clockwise.

1. Use the loop and junction rules to write down the three equations for Figure 4-12 that correspond to Equations (1), (2), and (3) derived above for Figure 4-11.

2. Show that if you make the substitution $I_2' = -I_2$, then the three equations you just derived can be rearranged algebraically so they are *exactly the same* as Equations (1), (2), and (3).

3. Measure the actual values of the two fixed resistors of 75 and 39 Ω and the two battery voltages with your multimeter. List the results below.

Measured voltage (emf) of the 6-V battery ε_1:_____

Measured voltage (emf) of the 1.5-V battery ε_2:_____

Measured resistance of the 75-Ω resistor R_1:_____

Measured resistance of the 39-Ω resistor R_3:_____

4. Carefully rewrite Equations (1), (2), and (3) with the appropriate *measured* (*not rated*) values for emf and resistances substituted into them. Use 100 Ω for the value of R_2 in your equations. You will be setting a variable resistor to that value soon.

5. Solve these three equations for the three unknown currents, I_1, I_2, and I_3 in amps. Show your calculations in the space below.

Question 3-2: Do your currents actually satisfy the equations? Show below by direct substitution.

Now you can verify your Kirchhoff's rules solution for this circuit. In addition to the materials from the previous activity, you will need:

- 0- to 200-Ω variable resistor (potentiometer)
- 6 alligator clip leads

Activity 3-2: Testing Kirchhoff's Rules with a Real Circuit

1. Using the resistance mode of the multimeter, measure the resistance between the center wire on the variable resistor and one of the other wires. What happens to the resistance reading as you rotate the dial on the variable resistor clockwise? Counterclockwise?

2. Set the variable resistor so that there is 100 Ω between the center wire and one of the other wires. Was it difficult to do?

3. Wire up the circuit pictured in Figure 4-11 using the 0- to 200-Ω variable resistor set at 100 Ω as R_2. Spread the wires and circuit elements out on the table so that the circuit looks as much like Figure 4-10 as possible.

4. Use the multimeter to measure the current in each branch of the circuit (see Note below), and enter your data in Table 4-3. Compare the measured values to those calculated in Activity 3-1 by computing the percent difference in each case.

> **Note:** The most accurate and easiest way to measure the currents with the digital multimeter is to measure the voltage across a resistor of known value, and then use Ohm's law to calculate I from V and R.

Table 4-3

	R measured w/ multimeter (Ω)	V measured w/ multimeter (V)	Measured $I = V/R$ (amps)	Theoretical I (amps) (from Activity 3-1)	% Difference
R_1					
R_2					
R_3					

Question 3-3: How well do your measured currents agree with the theoretical values you calculated in Activity 3-1? Are they within a few percent or do they differ by more than this?

If you have additional time, do the following extension.

Extension 3-3: How Do Changes Affect the Currents?

In Activities 3-1 and 3-2 you analyzed the circuit in Figure 4-11 with Kirchhoff's Loop Rule and Kirchhoff's Junction Rule. Now consider the case where resistor R_2 is removed from the circuit.

1. Draw a picture of the modified circuit (with R_2 removed) in the space to the right.

Question E3-4: Will the analysis of the circuit with Kirchhoff's rules now be more or less complex than it was for the circuit in Figure 4-10? That is, will you need to generate more or fewer equations with Kirchhoff's Loop Rule? With Kirchhoff's Junction Rule? Why?

Question E3-5: What is the effective value of R_2 in the modified circuit? Think very carefully and explain.

Prediction E3-1: Predict the values of the voltages across R_1 and R_3 for the modified circuit. Show your calculations clearly in the space below.

Test your prediction. Remove R_2 and measure I_1 and I_3 from V_1, V_3, R_1, and R_3 as before. Record the new values below.

V_1:_____ V_3:_____

I_1:_____ I_3:_____

Question E3-6: Did your observations agree with your predictions? Explain.

HOMEWORK FOR LAB 4
KIRCHHOFF'S CIRCUIT RULES

1. Find the equivalent resistance of the following network. (All resistances are in ohms.) Show your work below.

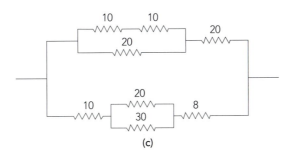

(c)

2. Show on the circuit diagram in Question 1 how you would connect a multimeter to measure the current through the 8-Ω resistor.

   a. Explain why the multimeter is connected in this way.

   b. What design feature of a good multimeter allows you to connect it in the way indicated without appreciably affecting the current through the 8-Ω resistor?

3. Show on the circuit diagram in Question 1 how you would connect a multimeter to measure the voltage across the 8-Ω resistor.

   a. Explain why the multimeter is connected in this way.

   b. What design feature of a good multimeter allows you to connect it in the way indicated without appreciably affecting the voltage across the 8-Ω resistor?

4. Pictured in the thick-lined box in the circuit on the right is a battery with emf 12 V and internal resistance 1 Ω. A is the positive terminal of the battery and B is the negative terminal.

   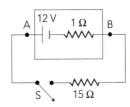

   a. What is the potential difference across the battery's terminals when the switch S is open, as shown?

b. What is the potential difference across the battery's terminals when the switch S is closed?

5. Determine the values and tolerances of resistors with the following color codes:

   a. red-red-brown-gold

   b. violet-gray-blue

   c. Could the value of a resistor marked as (A) actually be as large as 240 Ω? Explain.

6. Find the current through each of the resistors in the circuit on the right. Give both the magnitude and direction of current flow. Show all of your work below.

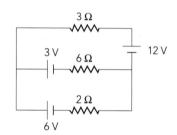

Name_____   Date_____

PRE-LAB PREPARATION SHEET FOR LAB 5— INTRODUCTION TO CAPACITORS AND RC CIRCUITS

(Due at the beginning of lab)

Directions:
Read over Lab 5 and then answer the following questions about the procedures.

1. Predict the change in capacitance of a parallel plate capacitor as the area of the plates is increased.

2. Predict the change in capacitance of a parallel plate capacitor as the separation between the plates is increased.

3. Briefly describe the observations you will make in Activity 1-2 to test one of these two predictions.

4. If you have two identical capacitors, what do you predict will be the capacitance of the two connected in parallel?

5. Briefly describe the observations you will make in Activity 2-1 to test this prediction.

6. Sketch below the complete circuits in Figure 5-5 with the switch in position 1, and with the switch in position 2.

7. What devices will you use to measure the decay of voltage in an RC circuit?

Lab 5:
Introduction to Capacitors and RC Circuits

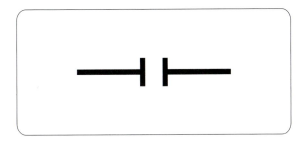

I get a real charge out of capacitors.

—P. W. Laws

OBJECTIVES

- To define capacitance and to learn to measure it with a digital multimeter.

- To discover how the capacitance of conducting parallel plates is related to the area of the plates and the separation between them.

- To explore and apply the rules for finding the equivalent capacitance of several capacitors connected in parallel and for several capacitors connected in series.

- To discover the effect of connecting a capacitor in a circuit in series with a resistor or bulb and a voltage source.

- To discover how the charge on a capacitor and the current through it change with time in a circuit containing a capacitor, a resistor, and a voltage source.

OVERVIEW

Capacitors are widely used in electronic circuits where it is important to store charge and/or energy or to trigger a timed electrical event. For example, circuits with capacitors are designed to do such diverse things as setting the flashing rate of Christmas lights, selecting what station a radio picks up, and storing electrical energy to run an electronic flash unit. Any pair of conductors that can be charged electrically so that one conductor has positive charge and the other conductor has an equal amount of negative charge on it is called a capacitor.

A capacitor can be made up of two arbitrarily shaped blobs of metal or it can have any number of regular symmetric shapes, such as one hollow metal sphere inside another, or a metal rod inside a hollow metal cylinder (Figure 5-1).

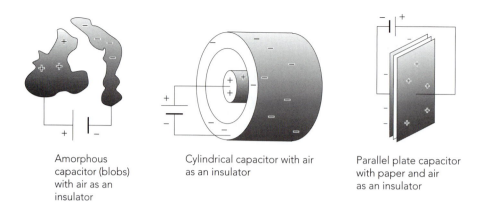

Amorphous
capacitor (blobs)
with air as an
insulator

Cylindrical capacitor with air
as an insulator

Parallel plate capacitor
with paper and air
as an insulator

Figure 5-1: Some different capacitor geometries.

The type of capacitor that is the easiest to analyze is the parallel plate capacitor. We will focus exclusively on the study of the properties of parallel-plate capacitors because the behavior of such capacitors can be predicted using only simple mathematical calculations and basic physical reasoning. Also, parallel plate-capacitors are easy to construct.

Although many of the most interesting properties of capacitors come in the operation of AC (alternating current) circuits (where current is first in one direction and then in the other), we will limit our present study to the behavior of capacitors in DC (direct current) circuits like those you have been constructing in the last couple of labs.

The circuit symbol for a capacitor is a simple pair of lines, as shown in Figure 5-2. Note that it is similar to the symbol for a battery, except that both parallel lines are the same length for the capacitor.

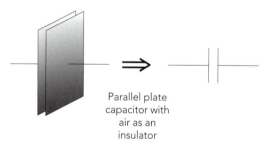

Parallel plate
capacitor with
air as an
insulator

Figure 5-2: The circuit diagram symbol for a
capacitor.

INVESTIGATION 1: CAPACITANCE, AREA, AND SEPARATION

The usual method for transferring equal and opposite charges to the plates of a capacitor is to use a battery or power supply to produce a potential difference between the two conductors. Electrons will then flow from one conductor (leaving a net positive charge) and to the other (making its net charge negative) until the potential difference produced between the two conductors is equal to that of the battery (see Figure 5-3).

In general, the amount of charge needed to produce a potential difference equal to that of the battery will depend on the size, shape, and location of the conductors relative to each other. (It will also depend on the properties of the material between the conductors.) The capacitance of a given capacitor is defined as the ratio of the magnitude of the charge, q (on either one of the conductors), to the voltage (potential difference), V, applied across the two conductors.

Thus,

$$C = q/V$$

Capacitance is defined as a measure of the amount of net or excess charge on either one of the conductors per unit potential difference. (The more charge a capacitor can store at a given voltage, the larger the capacitance.)

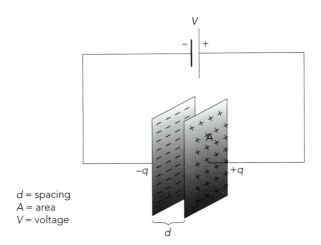

Figure 5-3: A parallel-plate capacitor with a voltage V across it.

You can draw on some of your experiences with electrostatics to think about what might happen to a parallel-plate capacitor when it is hooked to a battery, as depicted in Figure 5-3. This thinking can give you an intuitive feeling for the meaning of capacitance. For a fixed voltage from a battery, the net charge found on either plate is proportional to the capacitance of the pair of conductors and the applied voltage, $q = CV$.

Activity 1-1: Predicting the Dependence of Capacitance on Area and Separation

Consider two identical metal plates of area A that are separated by a distance d. The space between the plates is filled with a nonconducting material (air, for instance). Suppose each plate is connected to one of the terminals of a battery.

Question 1-1: What type of excess charge will build up on the metal plate that is attached to the negative terminal of the battery? What type of excess charge will build up on the plate that is connected to the positive terminal of the battery? Explain.

Question 1-2: Can the excess positive charges on one plate of a charged parallel-plate capacitor exert forces on the excess negative charges on the other plate? Explain.

Question 1-3: Use qualitative reasoning to anticipate how the amount of charge transferred to a pair of parallel-plate conductors (by connecting them to a battery) will change as the area A of the plates increases with the potential difference held constant. Explain your reasoning. How will this affect the capacitance of the capacitor? [**Hint:** Do the electric field and potential difference between the plates depend on the total charge on each plate or on the charge per unit area?]

Question 1-4: What happens to the potential difference between the plates of a parallel-plate capacitor when the separation d between the plates of the capacitor decreases while the excess charge is held constant? Explain. [**Hint:** What happens to the electric field between the plates as d is made smaller if the charge is kept constant by disconnecting the capacitor from the battery? What happens to the potential difference across the plates as d is made smaller after the capacitor is disconnected from the battery?]

The unit of capacitance is the farad, F, named after Michael Faraday. One farad is equal to one coulomb/volt. As you should be able to demonstrate to yourself shortly, the farad is a very large capacitance. Thus, actual capacitances are often expressed in smaller units with alternative notation as shown below:

microfarad: $1 \ \mu F = 10^{-6} \ F$

nanofarad: $1 \ nF = 1000 \ \mu\mu F = 10^{-9} \ F$

picofarad: $1 \ pF = 1 \ \mu\mu F = 1 \ UUF = 10^{-12} \ F$

(Note that M, m, μ, and U when written on a capacitor all stand for a multiplier of 10^{-6}.)

Several types of capacitors are typically used in electronic circuits, including disk capacitors, foil capacitors, and electrolytic capacitors. You should examine some typical capacitors. Your instructor will supply you with:

- a collection of old capacitors

To complete the next few activities you will need to construct a parallel-plate capacitor and use a multimeter to measure capacitance. Thus, you'll need the following items:

- 2 sheets of aluminum foil 8 × 8 inches
- pages in a "fat" textbook or phone book
- one or several bricks or other massive objects
- digital multimeter with a capacitance mode and clip leads
- ruler with a centimeter scale
- vernier calipers or a micrometer (optional)

- computer-based laboratory or data analysis software
- *RealTime Physics Electric Circuits* experiment configuration files

You can construct a parallel-plate capacitor out of two rectangular sheets of aluminum foil separated by pieces of paper. Pages in a textbook or phone book work well as the separator for the foil sheets. You can slip the two foil sheets between any number of pages, and weight the book down with something heavy like some bricks. The digital multimeter can easily be used to measure the capacitance of your capacitor.

Activity 1-2: Measuring How Capacitance Depends on Area or on Separation

Be sure that you understand how to set the meter and how to connect a capacitor to it. Devise a way to measure how the capacitance depends on *either* the foil area or on the separation between foils. Of course, you must keep the other variable (separation or area) constant.

When you measure the capacitance of your "parallel plates," be sure that the aluminum foil pieces are pressed together as uniformly as possible, and that they don't make electrical contact with each other.

If you hold the separation constant, record its value in Table 5-1. This may be measured in "pages," or the vernier caliper or micrometer may be used to translate this into meters. The area may be varied by using different size sheets of aluminum foil.

Alternatively, if you hold the area constant and vary separation, then record the dimensions of the foil so you will be able to calculate the area and enter it in Table 5-1.

1. Take at least five data points in either case. Record your data in Table 5-1.

Table 5-1

Separation (m)	Length (m)	Width (m)	Area (m^2)	Capacitance (nF)

Question 1-5: Describe how you measured all of the quantities in Table 5-1. Explain how you varied the separation or the area.

2. After you have collected all of your data, open the experiment file called **Dependence of C (L05A1-2)**. **Enter** your data for capacitance and either separation or area from Table 5-1 into the table in the software. Graph capacitance vs. either separation or area. **Print** the graph and affix it in the space below.

3. If your graph looks like a straight line, use the **fit routine** in the software to find its equation. If not, you should try other functional relationships until you find the best fit. **Print** and affix any other graphs above.

Question 1-6: What is the function that best describes the relationship between separation and capacitance *or* between area and capacitance? How do the results compare with your prediction based on physical reasoning?

Question 1-7: What difficulties did you encounter in making accurate measurements?

The actual mathematical expression for the capacitance of a parallel-plate capacitor of plate area A and plate separation d (Figure 5-3) is derived in your textbook. The result when there is vacuum (or just a low-density gas) between the plates is

$$C = \varepsilon_0 A/d$$

where $\varepsilon_0 = 8.85 \times 10^{-12}\ C^2/Nm^2$.

Question 1-8: Do your predictions and/or observations on the variation of capacitance with plate area and separation seem to agree qualitatively with this result? Explain.

Question 1-9: Use one of your actual areas and separations that corresponds to measurements you made in Activity 1-2 above to calculate a value of C using this equation. Show your calculations. How does the calculated value of C compare with your measured value? What might be wrong with the theoretical model for a capacitor in describing the behavior of your capacitor in this activity?

Question 1-10: In theory, what length and width in miles would big square foil sheets separated by a distance of 1 mm with wax paper have to be on each side for you to construct a 1-F capacitor? Show your calculations. Assume that wax paper has the same electrical properties as air. [**Hint:** Miles are not meters! In fact, 1000 m = 1 km = 0.62 mile.]

$L =$ _____ miles

INVESTIGATION 2: CAPACITORS IN SERIES AND PARALLEL

Take a look at an array of actual capacitors. They come in all sizes, shapes, and colors. You can measure their capacitances with the multimeter. You can also connect them in various series and parallel combinations and measure the equivalent capacitances of these combinations.

The definitions of series and parallel are the same as for other circuit elements like resistors (see Figure 5-4).

In a series connection, there is only one path for the charge. Whatever charge is placed on one of the capacitors must also be transferred to the other(s). In a parallel connection, the two terminals of each capacitor are connected directly to the terminals of the other(s). Each capacitor defines a branch, so that the total charge transferred to the capacitor combination is divided among the different capacitors.

To examine the equivalent capacitance of two capacitors connected in parallel or series, you'll need:

- 2 different capacitors (each about 0.1 μF)

- multimeter with a capacitance mode

- 6 alligator clip leads

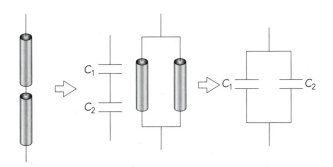

Figure 5-4: Capacitors wired in series and in parallel.

Activity 2-1: Equivalent Capacitance for Parallel Connection of Capacitors

Prediction 2-1: Use direct physical reasoning to predict the equivalent capacitance of a pair of capacitors with capacitance C_1 and C_2 wired in parallel. Explain your reasoning below. [**Hint:** What is the effective area of two parallel-plate capacitors wired in parallel? Does the effective separation between plates change when they are connected in this manner?]

1. Measure the capacitance of each capacitor with the multimeter.

   C_1:_____ C_2:_____

2. Connect the two capacitors in parallel and measure the equivalent capacitance of the parallel combination.

   C_{eq}:_____

Question 2-1: From your measurements, figure out a general equation for the equivalent capacitance of a parallel network in terms of C_1 and C_2. Explain how you reached your conclusion.

Question 2-2: How did your equation agree with your prediction? Explain.

Prediction 2-2: Use direct physical reasoning to predict the equivalent capacitance of a pair of capacitors wired in series. Explain your reasoning. [**Hint:** If you connect two capacitors in series what will happen to the charge along the conductor between them? What will the effective separation of the "plates" be? Will the effective area change?]

3. Connect the same two capacitors in series and measure the equivalent capacitance of the series combination.

   C_{eq}:————

Question 2-3: Are your measurements, consistent with the following equation for combining capacitors connected in series into an equivalent capacitance C_{eq}?

$$\frac{1}{C_{eq}} = \frac{1}{C_1} + \frac{1}{C_2}$$

Show the calculations you used to reach your conclusion.

Question 2-4: How does the equation given in Question 2-3 agree with your prediction? Explain.

INVESTIGATION 3: CHARGE BUILDUP AND DECAY IN CAPACITORS

Capacitors can be connected with other circuit elements. When they are connected in circuits with resistors, some interesting things happen. In this investigation you will explore what happens to the voltage across a capacitor when it is placed in series with a resistor in a direct current circuit. From your observations, you should be able to devise qualitative and quantitative explanations of what is happening.

   For the activities in this investigation you will need:

- computer-based laboratory system
- 2 current and 2 voltage probes
- *RealTime Physics Electric Circuits* experiment configuration files

- 6-V battery
- #133 flashlight bulb and socket
- 2 capacitors (about 25,000 μF)
- 6 alligator clip wires
- single-pole–double-throw switch
- 2 22-Ω resistors

You can first use a bulb in series with one of the amazing new ultracapacitors with a large capacitance. These will allow you to see what happens. Later on, to get more quantitative result, the bulb will be replaced by a resistor with a constant resistance.

Activity 3-1: Observations with a Capacitor, Battery, and Bulb

1. Set up the circuit shown in Figure 5-5. (If you are using a polar capacitor, be sure that the positive and negative terminals of the capacitor are connected correctly.)

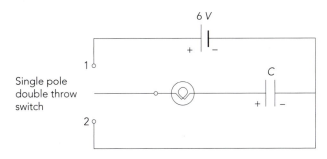

Figure 5-5: Circuit to examine the charging and discharging of a capacitor through a light bulb.

Question 3-1: Sketch the complete circuit for current when the switch is in position 1, and when it is in position 2.

Position 1   *Position 2*

2. Move the switch to position 2. After at least 30 s, switch it to position 1 and describe what happens to the brightness of the bulb.

Question 3-2: Draw a sketch on the axes below of the *approximate* brightness of the bulb as a function of time for the above case where you move the switch to position 1 after it has been in position 2 for a long time. Let $t = 0$ be the time when the switch was moved to position 1.

3. Now move the switch back to position 2. Describe what happens to the bulb. Did the bulb light again without the battery in the circuit?

Question 3-3: Draw a sketch on the axes below of the *approximate* brightness of the bulb as a function of time when it is placed across a charged capacitor *without the battery present*, i.e., when the switch is moved to position 2 after being in position 1 for a long time. Let $t = 0$ when the switch is moved to position 2.

Question 3-4: Can you explain why the bulb behaves in this way? Is there charge on the capacitor after the switch is in position 1 for awhile? What happens to this charge when the switch is moved back to position 2?

4. Open the experiment file called **Capacitor Decay (L05A3-1),** and display the axes that follow.

5. Connect the two probes to the interface, and **calibrate** them or **load the calibration**. **Zero** the probes with nothing attached to them.

6. Connect the probes in the circuit as in Figure 5-6 to measure the current through the light bulb and the potential difference across the capacitor.

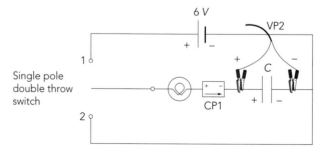

Figure 5-6: Current and voltage probes connected in Figure 5-5.

REALTIME PHYSICS: ACTIVE LEARNING LABORATORY

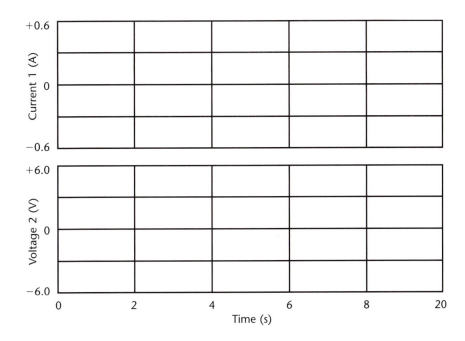

7. Move the switch to position 2. **Begin graphing.** When the graph lines appear, move the switch to position 1. When the current and voltage stop changing, move the switch back to position 2.

8. Sketch the graphs on the axes above, or **print** them and affix them over the axes.

9. Indicate on the graphs the times when the switch was moved from position 2 to position 1, and when it was moved back to position 2 again.

Question 3-5: Does the actual behavior over time observed on the current graph agree with your sketches in Questions 3-2 and 3-3? Do any features of the graphs surprise you? Explain.

Question 3-6: Based on the graph of potential difference across the capacitor, can you explain why the bulb lights when the switch is moved from position 1 to position 2 (when the bulb is connected to the capacitor with no battery in the circuit). Also explain the way the brightness of the bulb changes with time.

As you have seen before, a bulb does not have a constant resistance. Instead, its resistance is temperature dependent and goes up when it is heated by the current through it. For more quantitative studies of the behavior of a circuit with resistance and capacitance, you should replace the bulb with a 22-Ω resistor.

Activity 3-2: The Rise and Decay of Voltage in an RC Circuit

1. Replace the light bulb in your circuit (Figure 5-6) with a 22-Ω resistor. Move the switch to position 2. **Begin graphing**. When the graph lines appear, move the switch to position 1. When current and voltage stop changing, move the switch back to position 2.

2. Sketch your graphs on the axes that follow, or **print** them and affix them over the axes.

3. Indicate on the graphs the times when the switch was moved from position 2 to position 1, and when it was moved back to position 2 again.

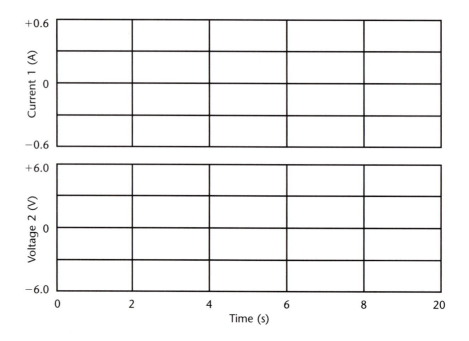

Question 3-7: Do the graphs for the resistor appear similar to those for the bulb? Are there any significant differences?

4. Use the **analysis feature** of the software to determine from your graph the *time constant* (the time for the voltage across the capacitor to decay to 37% of its initial value—after the switch is moved to position 2). Record your data below.

Initial voltage:_____   Time:_____

37% of initial voltage:_____   Time:_____

                                    Time constant:_____

If you made careful measurements of V vs. t for a capacitor C discharging through a resistor R, you should have gotten what is known as an *exponential decay curve*. This curve has exactly the same mathematical form as the cooling curve you may have encountered in the study of heat and temperature.

Mathematical reasoning based on the application of Ohm's law as well as the definitions of current and capacitance can be used to show that the following equation represents the voltage V across the capacitor as a function of time:

$$v(t) = V_0 \, e^{-t/RC}$$

In this equation, V_0 is the initial potential difference across the capacitor. (Note that V_0 is not necessarily the voltage of the battery.)

Question 3-8: Do the curves you measured for the decay of the potential difference across the capacitor in series with a resistor have the shape described by an exponential decay? Explain.

Question 3-9: Use the exponential function to calculate the *time constant* for your capacitor–resistor combination. [**Hint:** What time t in the function would make the value for $v(t)$ be just $0.37V_0$? (This is just V_0/e, where e is the base of natural logarithms.)] Use your values for R and C. Show all work. Does this value agree with your measured value?

Does this mathematical function describe the data you collected? If you have more time, do the following extension to examine this question.

Extension 3-3: Does the Observed Decay Curve Fit Theory?

1. Use the **fit routine** in the software to see if the voltage decay curve can be fit by an exponential function. Be sure to **select** and fit only that portion of the voltage graph where the voltage is decaying (decreasing)—up to where the voltage has *just reached* its minimum value.

2. **Print** the graph and affix it over the axes above, and record the equation of the function in Question 3-10.

Question E3-10: Did the exponential function fit your data well? Is the decay of voltage across the capacitor an exponential decay?

Question E3-11: Find the value of RC from the exponent in the function that fit the data. Compare this value to the one calculated from the resistance and capacitance. Do they agree? (Remember that the resistance and capacitance values are each known only to about $\pm 10\%$.)

If you have more time, do the following extension.

Extension 3-4 Decay with a Larger Resistance and Capacitance

Prediction E3-1: What do you predict will happen if the capacitor has a larger resistance in series with it? How will this affect the time constant?

Test your prediction.

1. Increase the resistance to twice its value. (Should you connect two 22-Ω resistors in series or in parallel?)

2. Start with the switch in position 2 and graph as before. Again measure the time constant from your graph. Record your data below.

    Initial voltage:_____   Time:_____

    37% of initial voltage:_____   Time:_____

                                             Time constant:_____

Question E3-12:   What happened to the time constant? Did this agree with your prediction?

Prediction E3-2:   What will happen if the capacitance of the capacitor is larger? How will this affect the time constant?

Test your prediction.

3. Increase the capacitance to twice its value. (Should you connect two capacitors in series or in parallel?)

4. Start with the switch in position 2 and graph as before. Again measure the time constant from the graph. Record your data below.

    Initial voltage:_____   Time:_____

    37% of initial voltage:_____   Time:_____

                                             Time constant:_____

Question E3-13:   What happened to the time constant? Did this agree with your prediction?

HOMEWORK FOR LAB 5
INTRODUCTION TO CAPACITORS AND RC CIRCUITS

1. Explain in terms of the charge, electric field, and potential difference the dependence of capacitance of a parallel-plate capacitor on area and separation of the plates in the equation $C = \varepsilon_0 A/d$.

2. A 1.5-V battery is connected to a 250-μF capacitor. How much charge is found on the capacitor plates?

3. For the circuit on the right with two capacitors of different capacitance in series, indicate whether the statements below are TRUE or FALSE, and for each false statement, write a correct one.

   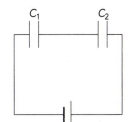

   a. Both capacitors have the same amount of charge on their plates.

   b. The voltages across both capacitors are the same.

   c. The sum of the voltages on the two capacitors equals the voltage of the battery.

4. Find the equivalent capacitance of each of the following combinations of capacitors. (All capacitances are in μF.)

   a.
   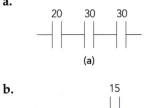
   (a)

   b.

   (b)

   c.
   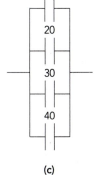
   (c)

5. In the circuit on the right, the capacitor is initially uncharged.

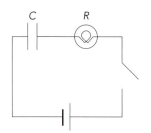

   a. Describe what is observed when the switch is closed.

   b. How would your observations be changed if the capacitor were twice as large?

   c. How would your observations be changed if the bulb had half as much resistance?

6. Sketch a graph of the current as a function of time in the circuit in Question 5 after the switch is closed. Also sketch a graph of the voltage across the capacitor as a function of time.

7. In the circuit on the right, the capacitor is initially charged. It has capacitance 0.023 F, while the resistor has resistance 47 Ω. How long after the switch is closed does the voltage on the capacitor fall to 37% of its initial value? Show your work.

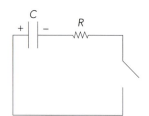

PRE-LAB PREPARATION SHEET FOR LAB 6—
INTRODUCTION TO INDUCTORS AND LR CIRCUITS
(Due at the beginning of lab)

Directions:
Read over Lab 6 and then answer the following questions about the procedures.

1. Why does an inductor have resistance?

2. What is the function of the switch in Figure 6-2? How will you use this circuit?

3. Sketch your answer to Question 2-2 in the box on the right.

4. Why do we examine the behavior of an inductor when the current through it is changing?

5. What is the definition of the time constant?

LAB 6:
INTRODUCTION TO INDUCTORS AND
LR CIRCUITS

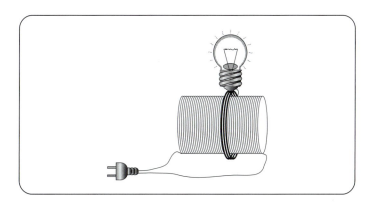

The power which electricity of tension possesses of causing an opposite electrical state in its vicinity has been expressed by the general term Induction . . .

—Michael Faraday

OBJECTIVES

- To examine the properties of a new circuit element—an inductor.

- To explore the behavior of a circuit that has a constant (DC) voltage source in series with an inductor and a resistor.

- To explore the behavior of a circuit that has a changing voltage source in series with an inductor and a resistor.

- To examine the time dependence of the current in a circuit with an inductor and a resistor.

OVERVIEW

You have learned that DC circuits with resistors have predictable voltages and currents. Ohm's law, $V = IR$, describes the proportional relationship between voltage and current for a resistor. Furthermore, you have discovered that resistors and capacitors in RC circuits also yield predictable currents and voltages. However, the relationship between voltage and current for an RC circuit is not proportional and varies in time. For example, the voltage drop across a discharging capacitor in an RC circuit obeys the relationship

$$v(t) = V_0 e^{-t/RC}$$

an exponential decay over time.

This lab will introduce you to the *inductor* and its role in a circuit. An inductor is a coil of wire. It responds to voltage and current *changes* in a different way than a capacitor does. While a capacitor responds to changes in the voltage across its terminals, an inductor responds to changes in the current through it.

In Investigation 1, you will examine the behavior of an inductor in a circuit in which the applied voltage is steady (DC). Then, in Investigation 2, you will look quantitatively at the behavior of an inductor in a circuit in which the applied voltage varies over time. In Investigation 3, you will examine the mathematical relationships between current through an inductor, the induced voltage across it, and time.

INVESTIGATION 1: DC BEHAVIOR OF AN INDUCTOR

In previous labs, we approximated the resistance of short wires as zero. We justified this approximation because the resistance of short wires is negligible compared to that of other circuit elements, such as resistors and light bulbs. As you may recall from your textbook, the resistance of a conducting wire increases with length. Thus, for a very long wire, the resistance may not be negligible.

An inductor is usually a long wire wound into a tight coil of circular loops. Consequently, all real inductors have resistance that is related to the length and type of wire used to wind the coil.

If you know what material a wire of length L is made of, you can calculate its resistance by looking up the experimentally determined resistivity ρ and using the relation $R = \rho L/A$, where A is the cross-sectional area of the wire.

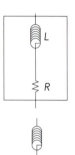

An *ideal* inductor has the properties of a real inductor, except it has no resistance. We can model a real inductor as an ideal inductor, L, in series with a resistor, R. (See the figure to the right.)

For simplicity, we will use the symbol on the right alone to represent a *real* inductor, bearing in mind that a real inductor always has some resistance.

In this Investigation you will need the following materials:

- computer-based laboratory system

- voltage probe and current probe

- *RealTime Physics Electric Circuits* experiment configuration files

- inductor (large coil of wire, approximately 3400 turns, 800 mH and about 60 Ω)

- 6-V battery

- 2   75-Ω resistors

- 7 alligator clip leads

- contact switch

- digital multimeter

In the next activity you will determine what happens over time when you connect a steady (DC) voltage source in series with a resistor and an inductor (real—with known resistance). In particular, you will examine the voltage across the inductor.

Activity 1-1: Inductors and Constant (DC) Voltages—Resistance Associated with an Inductor

1. Measure the resistance of your inductor with the digital multimeter. [**Note:** Your inductor's resistance may already be stamped on it—don't believe that value!]

    Resistance of inductor:_____

Now, consider the circuit in Figure 6-1.

Prediction 1-1: Predict the voltage across the inductor, and current through it, after the switch, S, has been closed for awhile. Be quantitative. Use the values of your circuit elements: $R_1 = 75\ \Omega$, $L = 800$ mH, R of the inductor (just measured), and $V = 6.0$ V. Record your predictions below. Clearly state your reasoning.

Predictions: Current:_____   Voltage:_____

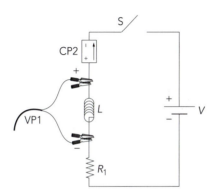

Figure 6-1: LR circuit with DC voltage applied.

2. Open the experiment file called **LR DC Circuit (L06A1-1).**

3. **Calibrate** the probes or **load the calibration. Zero** the probes with them disconnected from the circuit.

4. Connect the circuit shown in Figure 6-1, and measure the voltage across the inductor (VP1) and the current through it (CP2) after the switch is closed for awhile.

    Current:_____   Voltage:_____

Question 1-1: Do the values you measured agree with Prediction 1-1? If not, try to explain why not.

Question 1-2: Would you conclude that an inductor behaves the same or different than a resistor when a DC voltage is applied to a circuit containing it? Explain.

In the next activity, you will examine the behavior of resistors in circuits with applied voltage switching on and off (a *transient* voltage). Later, you will investigate the behavior of an inductor in a circuit with a switching voltage.

Consider the circuit in Figure 6-2 where $R_1 = R_2 = 75 \, \Omega$.

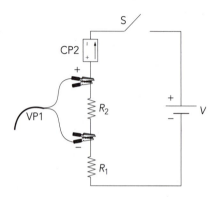

Figure 6-2: Two-resistor switching circuit.

Prediction 1-2: Two resistors of equal value are in series with a DC voltage source, with $V = 6.0$ V. You toggle switch S from open to closed to open several times, keeping it open or closed for a second or so each time. Predict the current through the circuit (CP2) and the voltage across R_2 (VP1) when the switch is open and closed. Be quantitative. Show your calculations below.

Switch Open: Current:_____   Voltage:_____

Switch Closed: Current:_____   Voltage:_____

Test your predictions.

Activity 1-2: Resistors in a Switching Circuit

1. Open the experiment file called **Switching Circuit (L06A1-2),** to display the axes that follow.

2. **Calibrate** the probes or **load the calibration,** if this hasn't already been done. **Zero** the probes with them disconnected from the circuit.

3. Connect the circuit in Figure 6-2.

4. **Graph** the current and voltage as you open and close the switch, keeping it open or closed for one second each time.

   Switch Open: Current:_____   Voltage:_____

   Switch Closed: Current:_____   Voltage:_____

5. **Adjust** the display of data so that the graphs are **persistently displayed.**

Question 1-3: Do your measurements agree with your predictions? If not, explain possible sources of discrepancy.

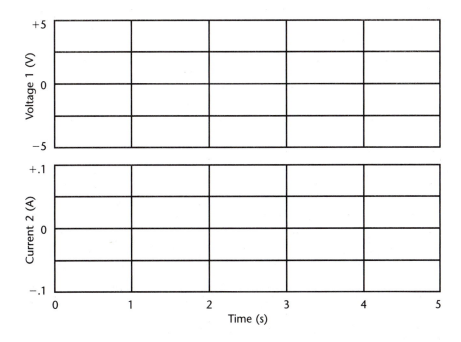

In Activity 1-1 you discovered that an inductor has resistance and behaves exactly like a resistor if a DC voltage is applied to it. In Activity 1-2 you observed the behavior of a resistor when the current through it changes. In the next activity, you will study the behavior of an *inductor* when the current through it changes.

Consider the circuit shown in Figure 6-3, in which an inductor is connected in series with a battery, resistor and switch.

Prediction 1-3: You have observed that resistors and inductors behave similarly in DC circuits. You have also observed the behavior of resistors in switching (transient) circuits. Do you think that an inductor will exhibit behavior similar to a resistor in a switching circuit? How might these two circuit elements differ? [**Hint:** If you have already studied Faraday's law, use it to predict your answer.]

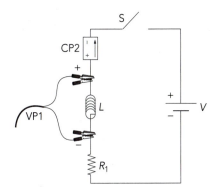

Figure 6-3: Switching circuit with inductor and resistor.

Test your predictions.

Activity 1-3: Inductors in Switching Circuits

1. Use the same experiment file, **Switching Circuit (L06A1-2)** and calibration as in Activity 1-2. Keep the graphs for the resistor from Activity 1-2 displayed.

2. **Zero** the probes with them disconnected from the circuit.

3. Connect the circuit in Figure 6-3.

4. **Graph** the current and the voltage as you toggle the switch, S, from open to closed. Be sure you keep it open or closed for approximately one second each time. [**Note:** The computer may not display the data in real-time. Be patient.]

5. **Print** your graphs for this activity and Activity 1-2, and affix over the previous axes. Label this graph LR, and the one from Activity 1-2 RR.

6. Mark on each set of graphs the time intervals when the switch was *open* and those when the switch was *closed*.

Question 1-4: Describe the differences and similarities between the LR and RR graphs. In particular, consider the time intervals when the current through the circuit is changing. Describe differences in the voltage across the inductor in the LR circuit and across the resistor, R_2, in the RR circuit.

Question 1-5: At the moment when you open the switch and the current drops to zero, what is the polarity of the voltage across the inductor. How is this related to the direction of the current *just before* you opened the switch?

Comment: As you saw in Activity 1-1, the DC behavior of an inductor is identical to that of a resistor. When there is a current through an inductor there is a voltage drop across the inductor because of the resistance of the wire. The voltage drop is proportional to the current, i.e., $V = IR$.

When you change the applied voltage, as in the switching circuit in Activities 1-2 and 1-3, the inductor exhibits a unique behavior, different from that of a resistor. It creates its own potential difference—an induced voltage. The polarity of the induced voltage is such that it opposes the change in the current through the inductor. As a result, it is impossible for you to instantaneously change the current through an inductor.

INVESTIGATION 2: INDUCTORS AND SELF-INDUCTANCE

In this investigation you will extend your observations of the behavior of inductors in circuits in which the current is changing.

In the next activity, you will predict the polarity of the induced voltage across an inductor in the LR circuit of Figure 6-4 at an instant *after* you disconnect or connect the applied voltage by opening or closing a switch. For the following questions and predictions, assume that the inductor, L, in this circuit is an *ideal* inductor, i.e., it has no resistance.

The circuit is similar to Figure 6-3. The only difference is an extra wire and switch S_2.

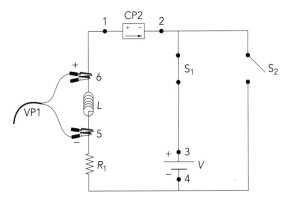

Figure 6-4: LR circuit.

The purpose of switch S_1 is to prevent the battery from wearing out when you are not collecting data. For the following discussions, assume switch S_1 is closed. However, *you should always open switch S_1 when you are not collecting data.*

The purpose of switch S_2 is to remove the battery as an energy source for the inductor while still providing a closed circuit for current flow through the inductor.

Note: Switch S_2 shorts out the battery. *Remember that S_1 and S_2 should be closed only while you are making measurements.*

Question 2-1: When the switch S_2 is closed, will any of the current produced by the battery go through CP2? Why or why not? [**Hint:** The resistance of the wire and switch in the branch containing S_2 is essentially zero.]

Question 2-2: The diagram on the left in Figure 6-5 depicts the equivalent circuit for Figure 6-4 when switch S_2 is open (with switch S_1 closed). In the space on the right in Figure 6-5, draw the equivalent circuit when switch S_2 is closed (with switch S_1 also closed). [**Hint:** Remember that the branch containing S_2 has zero resistance.]

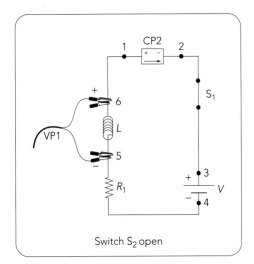

Switch S_2 open

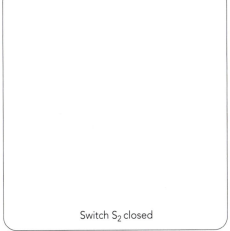

Switch S_2 closed

Figure 6-5: The left diagram is equivalent to Figure 6-4 with switch S_2 open and switch S_1 closed.

Question 2-3: If S_2 in Figure 6-4 has been open for a long time (with S_1 closed), what is the value of the current at points 1, 2, 3, 4, 5, and 6? Be quantitative. Use $R_1 = 75\ \Omega$, $L = 800$ mH, and $V = 6.0$ V (or the actual values of your circuit elements), and use the resistance of the inductor coil that you measured in Activity 1-1. Explain your calculations. [**Hint:** Use the diagram on the left in Figure 6-5.]

Question 2-4: If switch S_2 in Figure 6-4 has been open for a long time (with S_1 closed), what is the voltage (a) across the inductor, (b) across the resistor, (c) across the switch, S_2, and (d) across the battery? Again use the diagram on the left in Figure 6-5. Explain.

Predictions for the Voltage Across the Inductor Just After Switch S_2 Is Closed

Since the induced voltage across the inductor opposes any changes in the current, the current through the inductor just *after* you close S_2 must be the same as the current just *before* you close S_2. (If not, there would have to be an instantaneous change in current, which the inductor prevents from occurring.)

Prediction 2-1: Fill in the first column in Table 6-1 based on your answer to Question 2-3.

Prediction 2-2: What are the magnitude and direction of the current in the circuit just *after* you close S_2? Enter your predictions in the second column of Table 6-1.

Explain Predictions 2-1 and 2-2 in the space below Table 6-1.

Table 6-1

	S_2 has been open for a long time	*Immediately after S_2 is closed*	S_2 has been closed for a long time	*Immediately after S_2 is reopened*
Current in circuit: magnitude *and* (clockwise, zero, or counterclockwise)				
Induced voltage: ($V_{65} > 0$, $V_{65} = 0$, $V_{65} < 0$)				

Prediction 2-3: Based on your predictions for current in Table 6-1, will the potential difference V_{65} be greater than, less than, or equal to zero *immediately* after S_2 is closed? (That is, will the top end of the inductor [point 6] be at a higher

potential, the same potential, or a lower potential than the bottom end of the inductor [point 5]?) Write your prediction in Table 6-1 and explain below.

Question 2-5: If you keep S_2 closed for a long time, what is the current in Figure 6-4 at points 1, 2, 3, 4, 5, and 6? [**Hint:** Use the diagram on the right in Figure 6-5.] Explain.

Question 2-6: If you keep S_2 closed for a long time, what is the voltage (a) across the inductor, (b) across the resistor, and (c) across the switch? Explain.

Predictions for the Voltage Across the Inductor Just After Switch S_2 Is Opened:

Prediction 2-4: Fill in the third column in Table 6-1 based on your answer to Question 2-5.

Prediction 2-5: What are the magnitude and direction of the current in the circuit just *after* you open S_2. Enter your predictions in the last column of Table 6-1.

Prediction 2-6: Based on your predictions for current in Table 6-1, will the potential difference V_{65} be greater than, less than, or equal to zero *immediately* after S_2 is opened? (That is, will the top end of the inductor [point 6] be at a higher potential, the same potential, or a lower potential than the bottom end of the inductor [point 5]?) Write your prediction Table 6-1 and explain below.

Note: While analyzing the inductor, you considered *two* settings for S_2—open and closed. However, we must consider *four* time intervals. Table 6-2 lists two of these. Complete the table with the other two. (Remember, S_1 is always closed.)

Table 6-2

Time instance	Description
1	S_2 open for a long time
2	Just after S_2 is closed
3	
4	

Prediction 2-7: On the axes that follow, sketch your qualitative predictions for the induced voltage across the inductor and current through the inductor, for each of the four time intervals in Table 6-2. [**Hint:** Recall that the voltage across an inductor *can* change almost instantaneously, but the current through the inductor

cannot change instantaneously. Also, the induced voltage always opposes the change in current.]

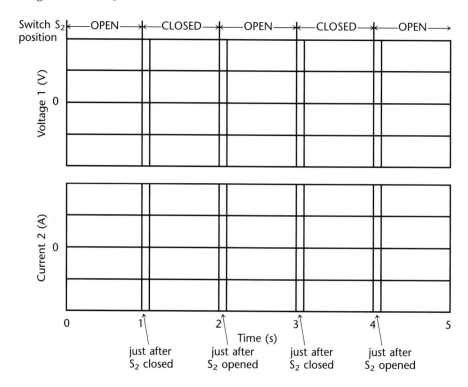

To test your predictions you will need the following.

- computer-based laboratory system
- voltage probe and current probe
- *RealTime Physics Electric Circuits* experiment configuration files
- large coil of wire (inductor) (approximately 3400 turns, 800 mH and about 60 Ω—same as in Investigation 1)
- 6-V battery
- 75-Ω resistor
- 7 alligator clip leads
- SPST knife switch
- contact switch

Activity 2-1: Observing the Polarity of the Induced Voltage Across an Inductor

1. Open the experiment file called **Inductor Polarity (L06A2-1).**

2. **Calibrate** the probes, or **load the calibration** if this hasn't already been done. **Zero** the probes while disconnected from the circuit.

3. Connect the circuit shown in Figure 6-4. Use the knife switch for S_1 and the contact switch for S_2. R_1 should be 75 Ω.

4. Close switch S_1 and leave it closed for the rest of this step. **Begin graphing** and measure the current through the inductor, CP2, and voltage across the

inductor, VP1, for the four time intervals listed in Table 6-2 by switching S_2 open and closed several times. You should hold each switch position for approximately one second.

5. **Print** your graphs and affix them over the axes above. *After you have collected your data, remember to open switch S_1 to prevent the battery from burning out.*

Question 2-7: Do your measurements agree with your predictions? If not, explain how they differ and offer some explanation for these differences.

Questions 2-8: Describe your observed evidence that the voltage induced across an inductor opposes any change in the current through the inductor.

INVESTIGATION 3: RL CIRCUITS

Using Faraday's law and the relationship between magnetic field and current for a coil, it can be shown that the induced voltage across an inductor is related to the time rate of change of the current through the inductor. The quantitative relationship is

$$V = -L \frac{\Delta i}{\Delta t} \tag{3–1}$$

In the above expression, V is the induced voltage across the inductor, $\Delta i/\Delta t$ is the time rate of change of current through the inductor, and L is the proportionality constant that relates the two. L is called the *inductance* of the inductor.

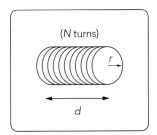

The value of L is a function of the number N of loops in the coil, the cross-sectional radius r of the coil, and the length d of the coil. For a closely wound cylindrical coil, similar to the inductor you have been using, the relationship is

$$L = \frac{\mu_0 N^2 \pi r^2}{d} \tag{3–2}$$

The constant μ_0 (called permeability of free space) has the value $4\pi \times 10^{-7}\,\text{T} \cdot \text{m/A}$. L has units of $\text{V} \cdot \text{s}^2/\text{C}$, called henries (H).

In Activity 2-2, you may have noticed that the current through an LR circuit decreased relatively slowly (compared to the RR circuit) when you toggled the switch S_2 from open to closed. The current also increased slowly to its maximum value, when you toggled the switch S_2 from closed to open.

There is a quantitative relationship for the current as a function of time after S_2 is closed. The relationship is

$$i(t) = I^{\max} e^{-t/(L/R)}$$

where $I^{max} = V/R$. In the above equation, $t = 0$ is the instant you closed S_2. As with the RC circuit in Lab 5, there is a time constant, τ, that characterizes the decays in an LR circuit. The time constant, $\tau = L/R$, describes the time it takes for the current to decrease to 37% of its original (maximum) value (I^{max}).

For the following questions and predictions, refer to the LR circuit in Figure 6-6. In this case, R is the resistance of the inductor coil, and there is no additional resistor connected in the circuit.

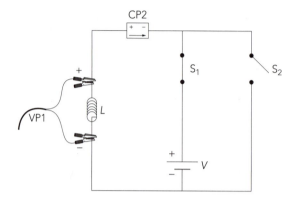

Figure 6-6: LR circuit for predictions and Activity 3-1.

Prediction 3-1: Calculate L for your inductor using Equation 3-2, and then calculate τ for the LR circuit in Figure 6-6. (Remember that R is the resistance of the coil, measured in Activity 1-1.) Record your predicted value of τ in the left-hand section of Table 6-3.

Prediction 3-2: What is the maximum current I^{max} through the inductor (the current when S_2 has been open for a long time)? Clearly state your reasoning. Show your calculations in the space below, and record your predicted values in the left-hand section of Table 6-3.

Prediction 3-3: Based on your calculation of I^{max}, what value should the current through the inductor have one time constant after the voltage is removed from the inductor (after S_2 is closed)? Show your calculation below, and enter your value as $I(\tau)$ in the left-hand section of Table 6-3.

To test your predictions by measuring the current before and after you close switch S_2, you will need the following:

- computer-based laboratory system
- voltage probe and current probe
- *RealTime Physics Electric Circuits* experiment configuration files
- large coil of wire (inductor) (approximately 3400 turns, 800 mH and about 60 Ω—same as in Investigation 1)
- 6-V battery
- 7 alligator clip leads

- SPST knife switch

- contact switch with very low noise or two alligator clip leads

Activity 3-1: Quantitative Analysis of Inductors and RL Circuit Time Constants

1. Open the experiment file called **RL Circuit (L06A3-1).** Axes similar to those that follow will be displayed. The software will be set up to trigger (begin graphing) when the voltage signal across the inductor decreases (when S_2 is closed).

2. **Calibrate** the probes, or **load the calibration** if this hasn't already been done. **Zero** the probes while disconnected from the circuit.

3. Connect the circuit shown in Figure 6-6. Use the knife switch for S_1 and the low noise contact switch (or alligator clip leads) for S_2. Remember, R is just the resistance of the inductor coil.

4. **Begin graphing** with S_2 open. After awhile, close S_2, and keep it closed until the graphs appear. Once graphing is finished, open S_1 and S_2.

5. After you collect the data, **adjust** the time axis on the current and voltage graphs if necessary to display about 0.3 seconds of data, including one complete transition from open to closed.

6. **Print** your graphs and affix them over the axes that follow.

7. Use the **analysis feature** to measure the maximum current I^{max} from your graph. Enter this value in Table 6-3.

8. Use the value from (7) to calculate the "measured" value of $I(\tau)$ and enter it in the table.

9. Use the value from (8) and the **analysis feature** in the software to accurately determine the time constant τ, the time it took for the current to drop to 37% of I^{max}. Record this value for τ in the table.

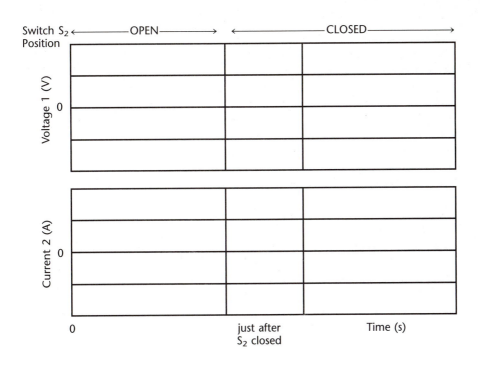

Table 6-3

	Inductor alone			Inductor and series resistor		
	I_{max} (mA)	$I(\tau)$ (mA)	τ (ms)	I_{max} (mA)	$I(\tau)$ (mA)	τ (ms)
Predicted						
Measured						

Question 3-1: Do your measurements agree with your predictions? If not, how do they differ?

Question 3-2: Describe the shape of your current vs. time graph just after you *close* the switch. Compare the shape of this curve to the ones you generated in Lab 5—RC circuits.

If you have more time, do the following extension. In addition to the equipment above, you will need:

• 75-Ω resistor

Extension 3-2: Decay with a Different Resistance

Prediction E3-3: If you add a 75-Ω resistor in series with the inductor and its resistance R in Figure 6-6, will the value of the time constant τ *increase, decrease,* or *stay the same*? Clearly explain your reasoning. If you predict that the value of τ will change, calculate the new value of τ, and enter this predicted value in the right hand section of Table 6-3. Show your calculation below.

Prediction E3-4: Calculate and fill in the predicted current values in the right-hand section of Table 6-3.

Add the 75-Ω resistor in series with the inductor. Reconnect the voltage probe so that it measures the voltage across the series combination of the inductor and resistor. Measure the value of the time constant τ of this circuit by repeating steps 4–9 in Activity 3-1. **Print** and label your graphs, and affix in the space below.

Question E3-3: Do your measurements agree with your predictions? If not, how do they differ?

Summary: In an LR circuit, an inductor will produce an induced voltage that opposes *changes* in the current through the circuit. The magnitude of this induced voltage is proportional to the time rate of change of current through the inductor. This induced voltage prevents rapidly changing current in the circuit.

There is a time constant, $\tau = L/R$, associated with the exponential decay of the current when the battery is shorted out of the circuit. In this expression, L is the inductance in henries, and R is the total resistance in ohms, including the resistance of the inductor coil. After one time constant, the current decays to 37% of its maximum value. (The same time constant applies to the exponential current buildup to its maximum value when the battery is connected to the circuit.)

The magnitude of the induced voltage also falls off exponentially from its maximum value at a rate that is related to the same time constant. After one time constant, the magnitude of the induced voltage has decayed to 37% of its maximum value.

HOMEWORK FOR LAB 6:
INTRODUCTION TO INDUCTORS AND LR CIRCUITS

1. You are asked to design an inductor of value 0.2 H. Specify the number of turns, cross-sectional radius, and length of the coil. Since there are an infinite number of possible responses, please explain what motivated your particular choice of parameters.

2. Suppose an inductor is constructed of copper wire of length 50 m that has a cross-sectional area of 7×10^{-7} m². (Copper has a resistivity of 2×10^{-8} Ω · m.) What is the resistance of this inductor?

3. Consider the circuit on the right. The switch has been open for a long time. (Assume the inductor is ideal; i.e., it has no resistance.)

   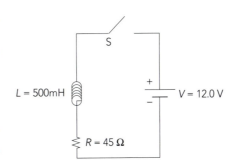

   a. What is the magnitude of the current through the inductor the *instant after* you close the switch? (Show all calculations and explain your answer.)

   b. Is the magnitude of the current constant, increasing, or decreasing as a function of time?

4. Consider the circuit in Question 3. Suppose you close switch S and let it remain closed for a very long time.

   a. What is the magnitude of the current through the inductor after you wait a very long time? (Show all calculations and explain your answer.)

   b. Is the magnitude of the current constant, increasing, or decreasing as a function of time?

5. Consider the circuit in Question 3. Suppose that the switch S has been closed for a very long time when you connect a wire across the battery, as shown in the figure on the right.

   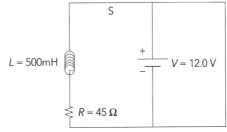

a. What is the magnitude of the current through the inductor the instant after you connect the wire? (Show all calculations and explain your answer.)

b. Is the magnitude of the current constant, increasing or decreasing as a function of time?

6. Consider the circuit in Question 3.

   a. What is the time constant τ for this circuit?

   b. What will be the magnitude of the current in the circuit τ seconds after you close the switch?

7. What is the time constant for a circuit composed of a real inductor ($L = 0.4$ H and $R = 50$ Ω) and a 150-Ω resistor connected in series?

8. You are asked to design an LR circuit with a time constant of 5×10^{-4} seconds. You have at your disposal a real inductor ($L = 0.02$ H, $R = 10$ Ω) and a wide variety of resistors. What value resistor should you choose for the circuit?

PRE-LAB PREPARATION SHEET FOR LAB 7—
INTRODUCTION TO AC CURRENTS AND VOLTAGES

(Due at the beginning of lab)

Directions:
Read Lab 7 and then answer the following questions.

1. What is a DC signal? What is an AC signal? Give examples of each.

2. If the unit for time is the second,

    a. What is the unit for frequency of a signal?

    b. What is the unit for period of a signal?

3. On the graph below, draw a sinusoidal signal with twice the frequency of the signal shown.

    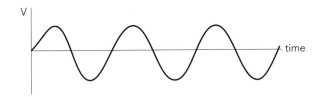

4. On the graph below, draw a sinusoidal signal that leads the signal shown by 90°.

    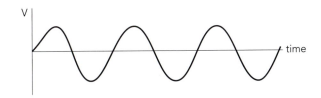

5. What does a signal generator do?

LAB 7:
INTRODUCTION TO AC CURRENTS AND VOLTAGES

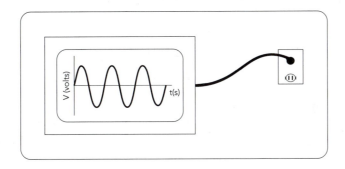

Electricity can be dangerous. My nephew tried to stick a penny into a plug. Whoever said a penny doesn't go far didn't see him shoot across that floor. I told him he was grounded.

—Tim Allen

OBJECTIVES

- To understand the difference between constant (DC) and time-varying (AC) current and voltage signals.

- To learn the meanings of *peak voltage* and *frequency* for AC signals.

- To observe the behavior of resistors for AC circuits.

- To observe the behaviors of capacitors and inductors for AC circuits.

- To understand the meanings of *phase, amplitude, reactance,* and *impedance* in AC circuits.

OVERVIEW

Until now, you have investigated electric circuits in which a battery provided an input voltage that was effectively constant in time. This is called a *DC* or *direct current* signal. (A steady voltage applied to a circuit eventually results in a steady current. Steady voltages are usually called *DC* voltages.)

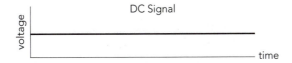

Signals that change over time exist all around you, and many of these signals change in a regular manner. For example, the electrical signals produced by your beating heart change continuously in time.

Examples of Time-Varying Signals

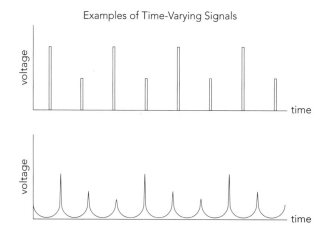

There is a special class of time-varying signals. These signals can be used to drive current in one direction in a circuit, then in the other direction, then back in the original direction, and so on. They are referred to as *AC* or *alternating current* signals.

Examples of AC Signals

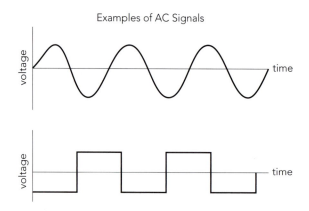

In Investigation 1, you will create an AC signal from the DC voltage of a battery. You will then learn about the signal generator—a device that produces a variety of regular, time-varying AC signals. In Investigation 2, you will discover how a time-varying signal affects a circuit with a resistor. In Investigation 3, you will discover how inductors and capacitors influence the current and voltage in various parts of an AC circuit.

INVESTIGATION 1: INTRODUCTION TO AC SIGNALS

The purpose of this investigation is for you to learn to use a battery and a signal generator to create time-varying signals. You will use a simple circuit consisting of a battery and a resistor, and will use probes to measure current and voltage in this circuit.

You will need the following materials:

- computer-based laboratory system

- current probe and voltage probe

- *RealTime Physics Electric Circuits* experiment configuration files

- 6-V battery

- 100-Ω resistor
- multimeter
- 7 alligator clip leads
- signal generator (50-Ω [LO- Ω] output impedance)

In the next activity, you will generate a time-varying signal from a battery's DC voltage.

Activity 1-1: Introduction to Time-varying Signals

Consider the circuit in Figure 7-1 consisting of a battery and a resistor, with current and voltage probes. In Figure 7-2a, the circuit diagram in Figure 7-1 is drawn for the case in which the wire labeled A is connected to the positive terminal of the battery, and the wire labeled B is connected to the negative terminal. (Let's call this *Position 1*.) The arrow labeled I represents the direction of current through the circuit (the flow of positive charge).

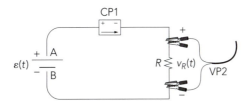

Figure 7-1: Circuit for generating a time-varying signal.

1. In the box for Figure 7-2b, sketch the circuit diagram for *Position 2*, in which you reverse the polarity. In other words, A is connected to the negative terminal of the battery, and B is connected to the positive terminal. Indicate the direction of current on your diagram with an arrow labeled I.

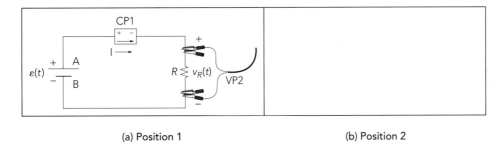

(a) Position 1   (b) Position 2

Figure 7-2: The circuit in Figure 7-1 drawn for the two possible connections to the battery.

Question 1-1: Describe the difference(s) between the circuits in Figure 7-2a and b.

Prediction 1-1: Suppose that you alternately connect the circuit in Position 1 and Position 2. If $R = 100 \; \Omega$ and $\varepsilon = 6.0$ V, sketch with dashed lines on the axes that follow your predictions for $i(t)$ and $v_R(t)$ when you briefly connect the circuit in

Position 1 for about a second, then reverse the connection to Position 2, then reverse it to Position 1 again, etc. Be quantitative—calculate the values of $i(t)$ and $v_R(t)$ for each position, and label the values of $i(t)$ and $v_R(t)$ on the vertical axes.

2. Open the experiment file **Time Varying Signal (L07A1-1)** to display the axes below.

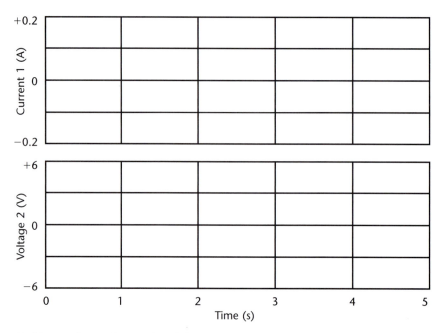

3. **Calibrate** the probes or **load the calibration,** and **zero** the probes with them disconnected from the circuit.

4. Connect the circuit in Figure 7-1. Do not connect the battery yet.

5. **Begin** graphing, then change the circuit from Position 1 to 2 and back at a rate of about two switches per second. (**Note:** it is easier to simply hold the alligator clips against the battery terminals rather than connect them.)

6. Sketch your graphs of current and voltage vs. time using *solid* lines on the axes above, or **print** your graphs and affix them over the axes.

Question 1-2: Do your measured graphs agree with your predicted ones? If not, in what ways do they differ and why?

Question 1-3: What is the peak value of $v_R(t)$? (**Note:** peak value is the same as amplitude.)

Question 1-4: Approximately, what is the frequency of the signal you generated? (**Note:** frequency is the number of *complete* cycles per second.) Explain how you determined the frequency.

Activity 1-2: Introduction to the Signal Generator

In the previous activity, you generated a time-varying signal by reversing the polarity of a circuit by hand. As you might imagine, it is difficult to generate a continuous AC signal at a constant frequency with this method. Instruments exist that can create time-varying signals with very regular behavior over time. Not surprisingly, such an instrument is called a *Signal Generator*.

1. Open the experiment file called **Signal Generator (L07A1-2).** This will set up the voltage probe and software to graph voltage vs. time in **Repeat Mode** on axes like those that follow.

2. **Load the calibration,** if this has not already been done, and **zero** the voltage probe with it disconnected from the signal generator.

3. Configure the signal generator to produce a sinusoidal voltage signal with amplitude of 3V (+3V maximum and −3V minimum) and frequency of 20 Hz (20 cycles/second).

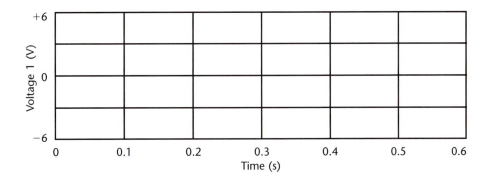

4. Connect the voltage probe across the signal generator.

5. **Begin graphing** and measuring the voltage output from the signal generator. Watch the graph for several seconds, and then **stop graphing** so that your measurements are captured.

6. **Print** your graph and affix it over the axes above.

Question 1-5: Compare the signal generated by the signal generator to the one you produced by hand in Activity 1-1. In what ways are they similar, and in what ways do they differ?

Question 1-6: Use the **analysis feature** to find the period (time from one peak to the next), *T*. (**Hint:** determine the time period for ten peaks and divide by 10.) Show your calculation.

$$T:\underline{\hspace{2cm}}s$$

Question 1-7: The period of a signal is the inverse of the frequency ($T = 1/f$). Is your measurement of period consistent with the frequency setting on the signal generator? Show your calculation.

7. **Begin graphing** again with the signal generator set at 20 Hz and 3 volts peak. While the voltage probe is continuously measuring, reduce the frequency of the signal generator to around 10 Hz, and then increase it to 40 Hz.

Question 1-8: Describe qualitatively how the displayed signal changes when you change the frequency from 20 Hz to 10 Hz, and from 10 Hz to 40 Hz. Discuss in terms of any changes in amplitude, frequency and period of the signal.

8. Change the signal generator frequency back to 20 Hz. **Begin graphing** again. While the voltage probe is continuously measuring the output signal, increase the amplitude of the voltage signal to 5 V (by watching your graphs) and then, after awhile, decrease it to 1 V.

Question 1-9: What changes in the displayed signal do you observe when you change the signal from 3 V to 5 V and then to 1 V? Discuss in terms of any changes in amplitude, frequency and period of the signal.

INVESTIGATION 2: AC SIGNALS AND RESISTANCE

In Investigation 1, you generated and observed the characteristics of a time-varying signal. You also learned how to operate a signal generator. In this investigation, you will consider the behavior of resistors in a circuit driven by AC signals of various frequencies.

You will need the following materials:

- computer-based laboratory system
- current probe and voltage probe
- *RealTime Physics Electric Circuits* experiment configuration files
- 100-Ω resistor
- multimeter
- 7 alligator clips leads
- signal generator (50 Ω [LO- Ω] output impedance)

Activity 2-1: Resistors and Time-Varying (AC) Signals

Consider the circuit with a signal generator and a resistor shown in Figure 7-3.

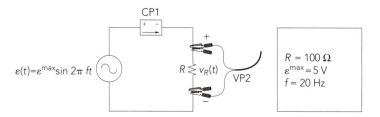

Figure 7-3: Resistor circuit with AC input signal.

Question 2-1:   What is the relationship between the input signal $\varepsilon(t)$ and the voltage $v_R(t)$ measured by the voltage probe? [**Hint:** Remember that CP1 has a very small resistance compared to R.]

Prediction 2-1:   On the axes that follow, sketch, with *dotted* lines, your *quantitative* predictions for the voltage $v_R(t)$ across R and the current $i(t)$ through R as functions of time. [**Hint:** Consider Ohm's law]. Label these graphs as Predictions.

Test your predictions.

1. Open the experiment file called **Resistor with AC (L07A2-1)** to display the axes that follow.

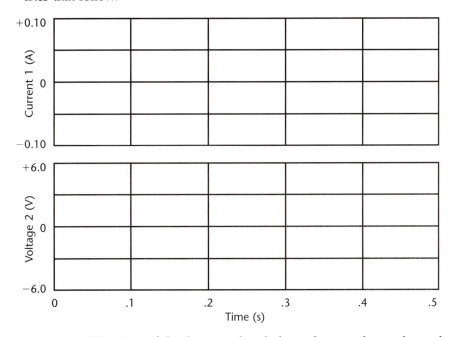

2. **Load the calibration,** if this has not already been done, and **zero** the probes with them disconnected from the circuit.

3. Connect the circuit in Figure 7-3.

4. Set the signal generator to 20 Hz and 5 V amplitude (+5 V maximum and −5 V minimum). [**Note:** You can check the amplitude now and during your observations by connecting the voltage probe temporarily directly to the signal generator.]

5. **Begin graphing.** When you have a good graph of the signal, **stop graphing.**

6. **Print** your graphs and affix them over the axes above.

7. On the graph of *voltage* vs. time, identify and label a time or two when the *current* (yes, the _current_) through the resistor is maximum.

8. On your graph of *current* vs. time, identify and label a time or two when the *voltage* (yes, the _voltage_) across the resistor is maximum.

Question 2-2:   Does a voltage maximum occur at the same time as a current maximum, or does one maximum (current or voltage) occur before the other? Explain.

9. Use your graph to complete column 1 in Table 7-1. To get information from the graph, you can use the **analysis feature.** Select several cycles by highlighting them, and then you can use the **statistics feature** to find the maximum values for the voltage and current.

Table 7-1

Column 1 $f = 20$ Hz	Column 2 $f = 30$ Hz	Column 3 $f = 40$ Hz

At maximum voltage, current is (*circle one*):
  maximum, minimum,
  zero and increasing,
  zero and decreasing,
  nonzero and increasing,
  nonzero and decreasing,
  other
 	
max. voltage (V_R^{max}) = ____
 	
max. current (I^{max}) = ____
 	
$R = V_R^{max}/I^{max}$ = ____
	At maximum voltage, current is (*circle one*):
	  maximum, minimum,
	  zero and increasing,
	  zero and decreasing,
	  nonzero and increasing,
	  nonzero and decreasing,
	  other
	
	max. voltage (V_R^{max}) = ____
	
	max. current (I^{max}) = ____
	
	$R = V_R^{max}/I^{max}$ = ____
		At maximum voltage, current is (*circle one*):
		  maximum, minimum,
		  zero and increasing,
		  zero and decreasing,
		  nonzero and increasing,
		  nonzero and decreasing,
		  other
		
		max. voltage (V_R^{max}) = ____
		
		max. current (I^{max}) = ____
		
		$R = V_R^{max}/I^{max}$ = ____

10. Set the frequency of the signal generator to 30 Hz. Check that the amplitude is still 5 V. **Graph** $i(t)$ and $v_R(t)$ as before. Use the **analysis feature** to complete Column 2 in Table 7-1.

11. Set the frequency of the signal generator to 40 Hz. Check that the amplitude is still 5 V. **Graph** $i(t)$ and $v_R(t)$ as before, and complete Column 3 in Table 7-1.

Question 2-3: Based on the calculations in Table 7-1, what can you say about the resistance of R at different frequencies? Does its value appear to increase, decrease, or stay the same as the frequency increases? Explain your answer.

Question 2-4: When the input signal is 30 or 40 Hz, does a maximum positive current through R occur before, after, or at the same time as the maximum positive voltage across R?

Note: Do *not* disconnect this circuit, you will be using a very similar one in Investigation 3.

Comment: In this investigation you discovered that the resistance of a resistor does not change when the frequency of the AC signal applied to it changes. In the Investigation 3, you will examine the behavior of capacitors and inductors with AC signals applied to them.

INVESTIGATION 3: AC SIGNALS WITH CAPACITORS AND INDUCTORS

In AC circuits, the behavior of circuit elements like inductors and capacitors is in some ways similar to the behavior of resistors. With a resistor in a DC circuit, the resistance determines how much current will flow when a voltage is applied, according to Ohm's law.

In AC circuits there is a quantity called *impedance* associated with each circuit element that acts like the resistance in Ohm's law. In fact, the peak voltage V^{max} is related to the peak current I^{max} by the relationship $I^{max} = V^{max}/Z$, where Z is the impedance.

For a resistor, the impedance is just the same as the resistance. You have seen in the last investigation that the impedance of a resistor does not change as the frequency of the applied signal changes. In this investigation you will learn about the impedance of a capacitor and an inductor and how these depend on the frequency of the applied AC signal.

You will need the following materials:

- computer-based laboratory system
- current probe and voltage probe
- *RealTime Physics Electric Circuits* experiment configuration files
- 100-Ω resistor
- multimeter
- 47-μF capacitor
- 800-mH inductor
- 7 alligator clip leads
- signal generator (50-Ω [LO-Ω] output impedance)

Activity 3-1: Capacitors and AC Signals

Does the impedance of a capacitor change when the frequency of the applied signal changes?

In this activity, you will investigate this question by measuring the behavior of a capacitor when signals of various frequencies are applied to it. Specifically, you will look at the amplitude and the phase of the current through and voltage across it. (See the Comment that follows for an introduction to phase.)

> **Comment:** When the peak current through and peak voltage across a circuit element always occur at the same instant, the current and voltage are said to be *in phase*. In Activity 2-1, you observed the AC current–voltage characteristics of a resistor. The current and voltage are *in phase* for a resistor.
>
> When the peak current occurs at a different instant than the peak voltage, there is a *phase difference*, or the current and voltage are said to be *out of phase*. The phase difference can be expressed in degrees, radians, or fractions of a period.

Consider the circuit shown in Figure 7-4.

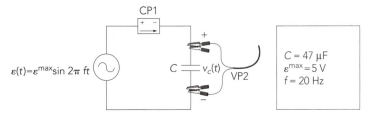

Figure 7-4: Capacitor circuit with AC input.

Prediction 3-1: Suppose that you replaced the signal generator with a battery and a switch. The capacitor is initially uncharged, and therefore the voltage across the capacitor is zero. If you close the switch, which quantity reaches its maximum value first: current in the circuit or voltage across the capacitor? As charge builds up on the capacitor, and the voltage across the capacitor increases, what happens to the current in the circuit? Explain.

Prediction 3-2: At the instant the capacitor reaches its maximum charge for this circuit, what do you predict the magnitude of the current will be—maximum, minimum, or zero? Why? At this instant, what must be the value of $v_C(t)$—maximum, minimum, or zero?

Prediction 3-3: The actual AC voltage applied to the circuit in Figure 7-4 by the signal generator is shown on the axes that follow. Use your answers from the above questions to sketch with *dotted* lines your prediction for the current as a function of time. Label your graph as prediction.

Test your predictions.

1. Open the experiment file called **Capacitor with AC (L07A3-1)** to display the axes that follow.

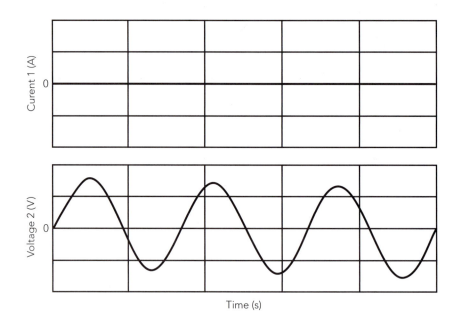

Time (s)

2. **Calibrate** the probes or **load the calibration,** if this has not already been done, and **zero** the probes while disconnected from the circuit.

3. Connect the circuit in Figure 7-4.

4. Set the signal generator to 20 Hz and amplitude of 5 V (+5-V maximum and −5-V minimum).

5. **Begin graphing.** When you have a good graph of the signal, **stop graphing.**

6. **Print** your graphs and affix them over the axes above.

7. On the graph of *voltage* vs. time, identify and label a time or two when the *current* (yes, the <u>current</u>) through the capacitor is maximum.

8. On your graph of *current* vs. time, identify and label a time or two when the *voltage* (yes, the <u>voltage</u>) across the capacitor is maximum.

9. Clearly mark one period of the AC signals on your graphs.

Question 3-1:   Does your measured current graph agree with your predicted one? If not, how do they differ?

Question 3-2:   For the capacitor with an input signal of 20 Hz, does a current maximum occur before, after, or at the same time as the maximum voltage? Explain.

> **Comment:**   One way you can determine the phase difference between two sinusoidal graphs with the same period is by measuring the time difference between neighboring peaks from each graph and dividing that time difference by the period. This will give you the phase difference as a fraction of a period. For example, if the time difference between the two peaks is 0.5 s and the period of the signals is 2.0 s, then the phase difference is 0.25 or 1/4 period. Phase difference can also be expressed in degrees or radians by simply multiplying the phase difference in periods by 360° or 2π. Continuing with the example, the signals are 90° or $\pi/2$ radians out of phase.
> The signal that reaches its peak value first is said to *lead* the other.
> The impedance X_C of a capacitor, is called the *capacitative reactance*. The relationship between the peak voltage (V_C^{max}) across the capacitor and peak current (I^{max}) through the capacitor is $V_C^{max} = I^{max} X_C$.

10. Use the analysis **feature** to help you fill in Column 1 in Table 7-2. Show your calculation of the phase difference in the space below the table.

Table 7-2

Column 1 $f = 20$ Hz	Column 2 $f = 30$ Hz	Column 3 $f = 40$ Hz
At maximum voltage, current is (*circle one*): maximum, minimum, zero and increasing, zero and decreasing, nonzero and increasing, nonzero and decreasing, other	**At maximum voltage, current is** (*circle one*): maximum, minimum, zero and increasing, zero and decreasing, nonzero and increasing, nonzero and decreasing, other	**At maximum voltage, current is** (*circle one*): maximum, minimum, zero and increasing, zero and decreasing, nonzero and increasing, nonzero and decreasing, other
max. voltage (V_C^{max}) = _____	max. voltage (V_C^{max}) = _____	max. voltage (V_C^{max}) = _____
max. current (I^{max}) = _____	max. current (I^{max}) = _____	max. current (I^{max}) = _____
$X_C = V_C^{max}/I^{max}$ = _____	$X_C = V_C^{max}/I^{max}$ = _____	$X_C = V_C^{max}/I^{max}$ = _____
Calculated phase difference: _____	Calculated phase difference: _____	Calculated phase difference: _____
Current leads or voltage leads?	Current leads or voltage leads?	Current leads or voltage leads?

11. Set the frequency of the signal generator to 30 Hz. Check that the amplitude is still 5 V. **Graph** $i(t)$ and $v_C(t)$ as before. Use the **analysis feature** to complete Column 2 in Table 7-2.

12. **Print** your graphs and affix them in the space below.

Question 3-3: What can you say about the magnitude of the reactance of the capacitor at 20 Hz compared to the reactance of the capacitor at 30 Hz? Explain based on your observations.

Question 3-4: What can you say about the phase difference between current and voltage for a capacitor at 20 Hz compared to the phase difference at 30 Hz? Explain based on your observations.

If you have time, complete the following extension.

Extension E3-2: Variation of Capacitative Reactance and Phase Angle with Frequency

Set the frequency of the signal generator to 40 Hz. Check that the amplitude is still 5 V. **Graph** $i(t)$ and $v_C(t)$ as before. Use the **analysis feature** to complete Column 3 in Table 7-2. **Print** your graphs and affix them below.

Question E3-5: What can you say about the magnitude of the reactance of the capacitor at 30 Hz compared to the reactance of the capacitor at 40 Hz?

Question E3-6: What can you say about the phase difference between current and voltage for a capacitor at 30 Hz compared to the phase difference at 40 Hz?

Question E3-7: Using the results from Table 7-2, suggest a mathematical relationship between your capacitor's reactance and the frequency of the signal. [**Hint:** A doubling in frequency had what effect on the reactance?]

Does the impedance of an inductor change when the frequency of the applied signal changes? Is there a phase difference between the current and voltage for an inductor? These questions will be answered in the following activity.

Activity 3-3: Inductors and AC Signals

Consider the circuit shown in Figure 7-5.

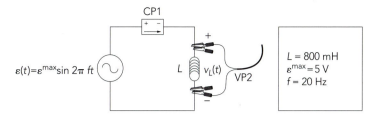

Figure 7-5: Inductor circuit with AC input.

Prediction 3-4: Suppose that you replace the signal generator with a battery and a switch. The inductor initially has no current through it. If you close the switch, which quantity reaches its maximum value first: current in the circuit or voltage across the inductor? [**Hint:** Recall from Lab 6 that when the current through an inductor is changing, the induced voltage across the inductor opposes the change.] As the current builds up in the circuit, what happens to the induced voltage across the inductor? Explain.

Prediction 3-5: At the instant the current reaches its maximum value for this circuit, what do you predict the magnitude of the induced voltage will be—maximum, minimum or zero? Why?

Prediction 3-6: The actual AC voltage applied to the circuit in Figure 7-5 by the signal generator is shown on the axes that follow. Use your answers from the above questions to sketch with *dotted* lines your prediction for the current as a function of time. Label your graph as Prediction.

Test your predictions.

1. Open the experiment file called **Inductor with AC (L07A3-3)** to display the axes that follow.

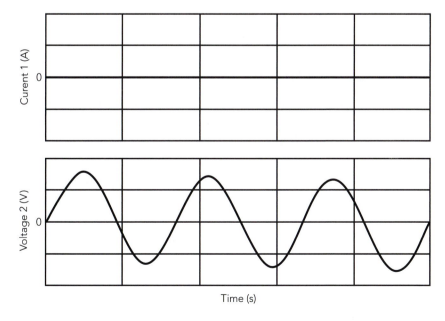

2. **Calibrate** the probes or **load the calibration,** if this has not already been done, and **zero** the probes while disconnected from the circuit.

3. Connect the circuit in Figure 7-5.

4. Set the signal generator to 20 Hz and amplitude of 5 V (+5 V maximum and −5 V minimum).

5. **Begin** graphing. When you have a good graph of the signal, **stop** graphing.

6. **Print** your graphs and affix them over the axes above.

7. On the graph of *voltage* vs. time, identify and label a time or two when the *current* (yes, the _current_) through the inductor is maximum.

8. On your graph of *current* vs. time, identify and label a time or two when the *voltage* (yes, the _voltage_) across the inductor is maximum.

9. Clearly mark one period of the AC signals on your graphs.

Question 3-8: Does your measured current graph agree with your predicted one? If not, how do they differ?

Question 3-9: For the inductor with an input signal of 20 Hz, does a current maximum occur before, after, or at the same time as the maximum voltage? Explain.

Comment: The impedance X_L of an inductor, is called the *inductive reactance*. The relationship between the peak voltage V_L^{max} across the inductor and peak current I^{max} through the inductor is $V_L^{max} = I^{max} X_L$. [**Note:** Since you are using a *real* inductor, it has some nonzero resistance as you measured in Lab 6. This will actually make your inductor act like a series combination of a resistor and ideal inductor.]

---

10. Use the **analysis feature** to fill in Column 1 in Table 7-3. Show your calculation of the phase difference in the space below the table.

Table 7-3

Column 1 f = 20 Hz	Column 2 f = 30 Hz	Column 3 f = 40 Hz
At maximum voltage, current is (*circle one*): maximum, minimum, zero and increasing, zero and decreasing, nonzero and increasing, nonzero and decreasing, other	**At maximum voltage, current is** (*circle one*): maximum, minimum, zero and increasing, zero and decreasing, nonzero and increasing, nonzero and decreasing, other	**At maximum voltage, current is** (*circle one*): maximum, minimum, zero and increasing, zero and decreasing, nonzero and increasing, nonzero and decreasing, other
max. voltage (V_L^{max}) = _____	max. voltage (V_L^{max}) = _____	max. voltage (V_L^{max}) = _____
max. current (I^{max}) = _____	max. current (I^{max}) = _____	max. current (I^{max}) = _____
$X_L = V_L^{max}/I^{max}$ = _____	$X_L = V_L^{max}/I^{max}$ = _____	$X_L = V_L^{max}/I^{max}$ = _____
Calculated phase difference: _____	Calculated phase difference: _____	Calculated phase difference: _____
Current leads or voltage leads?	Current leads or voltage leads?	Current leads or voltage leads?

11. Set the frequency of the signal generator to 30 Hz. Check that the amplitude is still 5 V. **Graph** $i(t)$ and $v_L(t)$ as before. Use the **analysis feature** to complete Column 2 in Table 7-3.

12. **Print** your graphs and affix them in the space below.

Question 3-10: What can you say about the magnitude of the reactance of the inductor at 20 Hz compared to the reactance at 30 Hz? Explain based on your observations.

Question 3-11: What can you say about the phase difference between current and voltage for a inductor at 20 Hz compared to the phase difference at 30 Hz? Explain based on your observations.

If you have time, complete the following extension.

Extension 3-4: Variation of Inductive Reactance and Phase Angle with Frequency

Set the frequency of the signal generator to 40 Hz. Check that the amplitude is still 5 V. **Graph** $i(t)$ and $v_L(t)$ as before. Use the **analysis feature** to complete Column 3 in Table 7-3. **Print** your graphs and affix them below.

Question E3-12: What can you say about the magnitude of the reactance of the inductor at 30 Hz compared to the reactance of the inductor at 40 Hz?

Question E3-13: What can you say about the phase difference between current and voltage for an inductor at 30 Hz compared to the phase difference at 40 Hz?

Question E3-14: Using the results from Table 7-3, suggest a mathematical relationship between your inductor's reactance and the frequency of the signal. [**Hint:** A doubling in frequency had what effect on the reactance?]

Summary: You discovered in Investigation 1 that the impedance (resistance) of a resistor does not change as the AC signal frequency is changed. The responses of a capacitor and an inductor differ from each other and from a resistor. As you have observed, as the frequency of the AC signal increases, the reactance of a capacitor decreases, while the reactance of an inductor increases. The following are the actual equations for capacitative and inductive reactance, respectively:

$$X_C = \frac{1}{2\pi f\, C} \quad \text{and} \quad X_L = 2\pi f\, L$$

Here, C is the capacitance, L is the inductance, and f is the frequency of the AC signal.

HOMEWORK FOR LAB 7: INTRODUCTION TO AC CURRENTS AND VOLTAGES

1. Consider a sinusoidal signal of frequency 20 Hz and amplitude 3 V displayed on a voltage vs. time graph. Describe qualitatively how the displayed signal changes when you change the frequency from 20 to 10 Hz. Answer in terms of the amplitude, frequency, and period of the AC signal.

2. Consider a sinusoidal signal of frequency 20 Hz and amplitude 3 V displayed on a voltage vs. time graph. Describe qualitatively how the displayed signal changes when you change the voltage from 3 to 6 V. Answer in terms of the amplitude, frequency, and period of the AC signal.

3. On the graph below, draw a sinusoidal current graph that *leads* the voltage already displayed by 90°. Label the curve you draw "current."

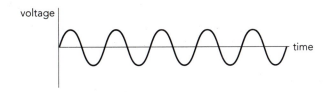

    If you generated these curves from circuits similar to ones in Investigations 2 and 3, which circuit device would produce this current when the displayed voltage is applied: a capacitor, inductor, or resistor? Explain based on your observations in this lab.

4. On the graph below, draw a sinusoidal current graph that *lags* the voltage already displayed by 90°. Label the curve you draw "current."

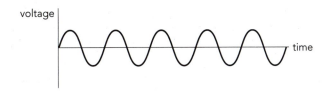

    If you generated these curves from circuits similar to ones in Investigations 2 and 3, which circuit device would produce this current when the displayed voltage is applied: a capacitor, inductor, or resistor? Explain based on your observations in this lab.

5. In the lab, you discovered that the impedance of a resistor (resistance) is independent of the frequency of the applied AC signal. On the graph below, qualitatively sketch the relationship between the impedance of a given resistor and the frequency of the applied signal. Be sure to label this curve *R*.

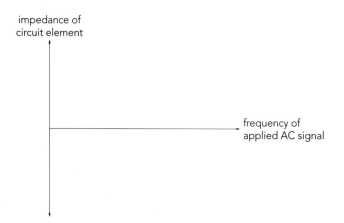

6. You also discovered that the impedance (reactance) of a capacitor is inversely proportional to the frequency of the applied AC signal. The reactance is also inversely proportional to the capacitance. On the axes above, qualitatively sketch the relationship between the impedance of a capacitor and frequency of the applied signal. Be sure to label this curve *C*.

7. Finally, you discovered that the impedance (reactance) of an inductor is proportional to the frequency of the applied AC signal. The reactance is also proportional to the inductance. On the axes above, qualitatively sketch the relationship between the reactance of a given inductor and frequency of the applied signal. Be sure to label this curve *L*.

PRE-LAB PREPARATION SHEET FOR LAB
8—INTRODUCTION TO AC FILTERS AND RESONANCE

(Due at the beginning of lab)

Directions: Read Lab 8 and then answer the following questions.

1. What does a capacitative filter do to an AC signal composed of many different frequencies?

2. List one application for filters and why they are useful in that application.

3. Describe in your own words what resonance means. Can you give a mechanical example of resonance?

4. The unit of resistance is the ohm.

   a. What is the unit of capacitance?

   b. What is the unit of capacitative reactance?

   c. What is the unit of inductance?

   d. What is the unit of inductive reactance?

LAB 8:
INTRODUCTION TO AC FILTERS AND RESONANCE

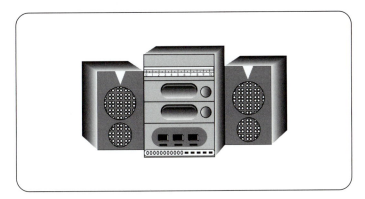

With electricity we were wired into a new world, for electricity brought the radio, a crystal set [and] with enough ingenuity, one could tickle the crystal with a cat's whisker and pick up anything.

—Theodore H. White

OBJECTIVES

• To understand the design of capacitative and inductive filters

• To understand resonance in circuits driven by AC signals

OVERVIEW

In Lab 7 you learned about alternating current (AC) signals that change over time. Such signals exist all around you. Many useful devices such as radio receivers and amplifiers, computers, and televisions use AC signals that are manipulated in precise ways. As you have seen in Lab 7, resistors, capacitors, and inductors have predictable effects on AC signals and consequently are important components in electronic devices.

In this lab you will continue your investigation of the behavior of resistors, capacitors, and inductors in the presence of AC signals. In Investigation 1, you will see how capacitors and inductors can act as "filters." More precisely, you will see how these elements can be used to suppress the voltage of certain frequency ranges of AC signals, while leaving other signals relatively unchanged.

In Investigation 2, you will discover the relationship between peak current and peak voltage for a series circuit composed of a resistor, inductor, and capacitor. You will also explore the phase difference between the current and the voltage. Furthermore, you will see how such a circuit can produce interesting responses, such as filtering out all but a narrow range of signal frequencies. This circuit is often referred to as a "resonant circuit." The phenomenon of resonance is a central concept underlying the tuning of a radio or television to a particular frequency.

INVESTIGATION 1: INTRODUCTION TO AC FILTERS (AND STEREO SPEAKERS)

In the previous lab, you explored the relationship between impedance (the AC equivalent of resistance) and frequency for a resistor, capacitor, and inductor. These relationships are very important to people designing electronic equipment, particularly audio equipment. You can predict many of the basic characteristics of simple audio circuits based on what you have learned in previous labs.

The purpose of this lab is for you to create circuits that filter out AC signals with frequencies outside the range of interest. In the context of these activities, a filter is a circuit that suppresses the voltage of some range of signal frequencies, while leaving other frequency ranges relatively unaffected.

You will need the following materials:

- computer-based laboratory system
- *RealTime Physics Electric Circuits* experiment configuration files
- 100-Ω resistor
- 47-μF capacitor
- 800-mH inductor
- 7 alligator clip leads
- signal generator (50 Ω [LO-Ω] output impedance)

Activity 1-1: Capacitors as Filters

In this activity, you will investigate how a circuit containing a resistor, capacitor, and signal generator responds to signals at various frequencies.

Consider the circuit in Figure 8-1 with a resistor, capacitor, signal generator and current and voltage probes.

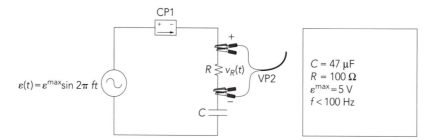

Figure 8-1: Capacitive filter circuit.

Prediction 1-1: On the left axes that follow, use dashed lines to sketch your *qualitative* prediction for the peak current I^{max} through the resistor as the *frequency of the signal from the signal generator is increased from zero*. [**Hint:** Recall that the capacitor's impedance is related to the frequency of the signal by the expression $X_C = 1/(2\pi f C)$.]

Prediction 1-2: On the right axes, sketch the peak voltage across the resistor V_R^{max} vs. *frequency.*

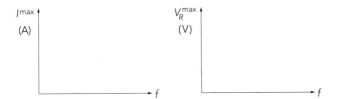

Explain how you arrived at your graphs. Discuss the relationship between the peak current and peak voltage.

Test your predictions.

1. Open the experiment file called **Capacitative Filter (L08A1-1).**

2. **Calibrate** the probes or **load the calibration. Zero** the probes while disconnected from the circuit.

3. Set the signal generator to a frequency less than 100 Hz and an amplitude of 5 V (+5 V maximum and −5 V minimum). **Note:** You can check the amplitude now and during your observations by connecting the voltage probe temporarily directly to the signal generator.

4. Connect the resistor, capacitor, signal generator, and probes as shown in Figure 8-1.

5. **Begin graphing.**

6. Use the **analysis feature** of the software to determine the peak voltage and peak current. (Use the statistics feature to find the maximum values for a number of cycles.) Enter your values in Table 8-1.

Table 8-1

$\varepsilon^{max}(V)$	$f(Hz)$	$V_R^{max}(V)$	$I^{max}(A)$

7. Change the frequency of the signal generator to another value less than 100 Hz, which is significantly different from the first. Be sure that the amplitude is still 5 V. (You should check this with the voltage probe.)

8. Repeat steps 5 and 6.

9. Repeat several more times with different frequencies. Be sure to enter at least 4–5 data points in Table 8-1.

10. Plot the data from Table 8-1 on the axes below. Mark scales on the vertical axes.

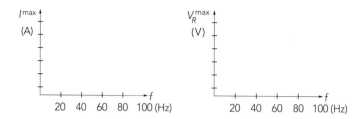

Question 1-1: If you could continue taking data up to very high frequencies, what would happen to the peak current I^{max} through the resistor and the peak voltage V_R^{max} across the resistor?

Question 1-2: At very high frequencies, does the capacitor act more like an open circuit (a break in the circuit's wiring) or more like a short circuit (a connection with very little resistance)? Justify your answer.

Question 1-3: What AC signal frequency does a DC signal have?

Prediction 1-3: What would be the current through and voltage across the resistor if you replaced the signal produced by the AC signal generator with a DC source?

Test your prediction by acquiring a DC data point.

11. Set the signal generator to its DC output mode. Be sure the DC signal has an amplitude of 5 V. (You should check this with the voltage probe.)

12. **Begin graphing.**

13. Use the **analysis feature** to determine the peak voltage and current.

14. Enter this data point in Table 8-1 and plot it on your graph.

Question 1-4: At very low frequencies, does the capacitor act more like an open circuit (a break in the circuit's wiring) or more like a short circuit (a connection with very little resistance)? Justify your answer.

Comment: In the last lab you learned that a capacitor's impedance, its *capacitative reactance*, decreases as the frequency of the AC signal increases and that the impedance of a resistor is independent of the signal frequency. Note that even though these elements are in series, their impedances do not simply add together. (This is because the current and voltage are not in phase.) The actual relation is: $Z = \sqrt{R^2 + X_C^2}$. Nevertheless, as you can see from this expression for Z, the impedance of the series circuit decreases as the reactance of the capacitor decreases. Since $X_C = 1/2\pi f C$, the impedance of the circuit decreases as the signal frequency increases.

The peak voltage and peak current are related to each other by a relationship that resembles Ohm's law: $V^{\max} = I^{\max}Z$.

In the circuit in Figure 8-1, since the peak voltage from the signal generator remains unchanged, the peak current in the circuit must increase as the total impedance decreases. Therefore, the peak voltage across the resistor, $V_R^{\max} = I^{\max}R$, increases as the frequency of the signal increases. This type of circuit is an example of a "high-pass" circuit or filter.

Activity 1-2: Inductors as Filters

This activity is very similar to the previous one except that you will replace the capacitor with an inductor and determine the filtering properties of this new circuit.

Consider the circuit containing a resistor, inductor, signal generator, and probes shown in Figure 8-2.

Prediction 1-4: On the left axes that follow, use dashed lines to sketch your *qualitative* prediction for the peak current $I^{\max}$ through the resistor as the *frequency of the signal from the signal generator is increased from zero.* [**Hint:** Recall that the inductor's impedance (inductive reactance) is related to the frequency of the signal by the expression $X_L = 2\pi f L$.]

Prediction 1-5: On the right axes, sketch the peak voltage $V^{\max}$ across the resistor vs. *frequency.*

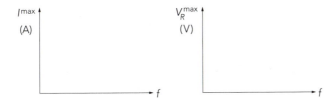

Explain how you arrived at your graphs. Discuss the relationship between the peak current and peak voltage.

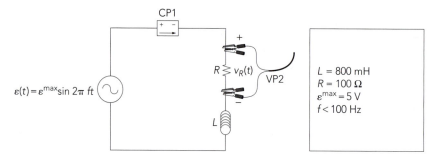

Figure 8-2: Inductive filter circuit.

Test your predictions.

1. Set the signal generator to a frequency less than 100 Hz and an amplitude of 5 V (+5 V maximum and −5 V minimum).

2. Open the experiment file called **Inductive Filter (L08A1–2).**

3. **Calibrate** the probes or **load the calibration** if this has not already been done. **Zero** the probes while disconnected from the circuit.

4. Connect the resistor, inductor, signal generator and probes as shown in Figure 8-2.

5. **Begin graphing.**

6. Use the **analysis feature** of the software to determine the peak voltage and peak current, and enter in Table 8-2.

Table 8-2

ε^{max} (V)	f (Hz)	V_R^{max} (V)	I^{max} (A)

7. Change the frequency of the signal generator to another value less than 100 Hz, which is significantly different from the first. Be sure that the amplitude is still 5 V. (You should check this with the voltage probe.)

8. Repeat steps 5 and 6.

9. Repeat several more times with different frequencies. Be sure to enter at least 4–5 data points in Table 8-2.

10. Plot the data from Table 8-2 on the axes below. Mark scales on the vertical axes.

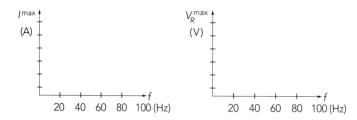

Question 1-5: If you could continue taking data up to very high frequencies, what would happen to the peak voltage V_R^{max} across the resistor and the peak current I^{max} through the resistor?

Question 1-6: At very high frequencies, does the inductor act more like an open circuit (a break in the circuit's wiring) or more like a short circuit (a connection with very little resistance)? Justify your answer.

Prediction 1-6: What would the current through and the voltage across the resistor be if you replaced the signal produced by the AC signal generator with a DC source?

Test your prediction by acquiring a DC data point.

11. Set the signal generator to its DC output mode. Be sure the DC signal has an amplitude of 5 V. (You should check this with the voltage probe.)

12. **Begin graphing.**

13. Use the **analysis feature** to determine the peak voltage and current.

14. Enter this data point in Table 8-2 and plot it on your graph.

Question 1-7: At very low frequencies, does the inductor act more like an open circuit (a break in the circuit's wiring) or more like a short circuit (a connection with very little resistance)? Justify your answer.

Comment: In the last lab you learned that an inductor's impedance, its *inductive reactance*, increases as the frequency of the AC signal increases and that the impedance of a resistor is independent of the signal frequency. As with the capacitor and resistor, the impedances of the inductor and resistor do not simply add together. (This is because the current and voltage are not in phase.) The actual relation is: $Z = \sqrt{R^2 + X_L^2}$. As you can see from this expression for Z, the impedance of the series circuit increases as the reactance of the inductor increases. Since $X_L = 2\pi f L$, the impedance of the circuit increases as the signal frequency increases.

Again, the peak voltage and peak current are related to each other by $V^{max} = I^{max}Z$.

In the circuit in Figure 8-2, since the peak voltage from the signal generator remains unchanged, the peak current in the circuit must decrease as the total impedance increases. Therefore, the peak voltage across the resistor, $V_R^{max} = I^{max}R$, decreases as the frequency of the signal increases. This type of circuit is an example of a "low-pass" circuit or filter.

Activity 1-3: Introduction to Audio Speaker Design

Music is sound waves composed of many different frequencies (pitches) superimposed on each other all at once and constantly changing. Low notes correspond to sound waves with a low frequency. High notes correspond to sound waves with a high frequency. A microphone is a device that can detect sound waves over a wide range of audible frequencies and convert them to AC signals. Thus, the signal from a microphone generally consists of many AC frequencies superimposed on each other all at once. (These signals can then be encoded and recorded on a storage device, such as a CD.)

Some of the ideas you have studied in this investigation are readily applied to the design of audio speakers. Many audio speakers have at least two separate "cones" from which sound emanates. Generally, the small cone, or tweeter, generates the high-frequency sound waves, and the large cone, or woofer, generates the low-frequency sound waves.

Speaker designers use cones of different size because small cones are generally better suited for playing the high notes and, conversely, big cones handle the low notes better. To best utilize the properties of the different sized cones, speaker designers use capacitative and inductive filter circuits to send high-frequency signals to the small cone and low-frequency signals to the large cone.

Imagine that you replace the signal generator in the circuits in Figure 8-1 or 8-2 with a collection of different frequency signals coming from your CD player. You also have a woofer and a tweeter. You can connect either across the resistor in one of the circuits, in place of the voltage probe.

Question 1-8: Into which circuit should you wire the woofer—the capacitative filter circuit or the inductive filter circuit? [**Note:** assume that the bigger the voltage across the resistor, the louder the sound emitted by the speaker.] Briefly explain your reasoning.

Question 1-9: Into which circuit should you wire the tweeter—the capacitative filter circuit or the inductive filter circuit? Briefly explain your reasoning.

How would one pass not high nor low, but middle frequency signals to a midsized cone, sometimes referred to as the "midrange"? You should find the answer to this question in Investigation 2.

INVESTIGATION 2: THE SERIES RLC RESONANT (TUNER) CIRCUIT

In this investigation, you will use your knowledge of the behavior of resistors, capacitors, and inductors in circuits driven by various AC signal frequencies to predict and then observe the behavior of a circuit with a resistor, capacitor, and inductor connected in series.

The RLC series circuit you will study in this investigation exhibits a "resonance" behavior that is useful for many familiar applications. One of the most familiar uses of such a circuit is as a tuner in a radio receiver.

You will need the following materials:

- computer-based laboratory system
- current probe and voltage probe
- *RealTime Physics Electric Circuits* experiment configuration files
- 100-Ω resistor
- 47-μF capacitor
- 800-mH inductor
- 7 alligator clip leads
- signal generator (50-Ω [LO-Ω] output impedance)

Consider the series RLC circuit shown in Figure 8-3.

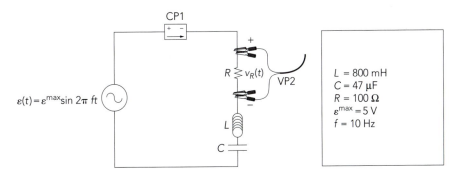

Figure 8-3: RLC series circuit.

Prediction 2-1: At very low signal frequencies (near 0 Hz), will the maximum values of $i(t)$ and $v_R(t)$ across the resistor be relatively large, intermediate, or small? Explain your reasoning.

Prediction 2-2: At very high signal frequencies (well above 100 Hz), will the maximum values of $i(t)$ and $v_R(t)$ be relatively large, intermediate, or small? Explain your reasoning.

Prediction 2-3: Based on your Predictions 2-1 and 2-2, is there some intermediate frequency where $i(t)$ and $v_R(t)$ will reach maximum or minimum values? Do you think they will be maximum or minimum?

Prediction 2-4: On the axes below, draw qualitative graphs of X_C vs. frequency and X_L vs. frequency. Clearly label each curve. [**Hint:** Base your answers on your observations in Lab 7 and Investigation 1 of this lab.]

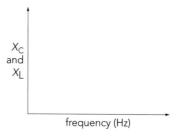

Comment: As we noted earlier in this lab, the relationship between the total impedance Z for a series combination of a resistor, capacitor, and inductor is not just the sum of the impedances of the three circuit elements. Instead, because of phase differences, Z is given by the following expression:

$$Z = \sqrt{R^2 + (X_L - X_C)^2}$$

Prediction 2-5: For what values of X_L and X_C will the total impedance of the circuit Z be a minimum? On the axes above, mark and label the frequency where this occurs. Explain your reasoning.

Prediction 2-6: At the frequency you labeled, will the value of the peak current I^{max} in the circuit be a maximum or minimum? What about the value of the peak voltage $V_R{}^{max}$ across the resistor? Explain your reasoning.

The point you identified for Predictions 2-5 and 2-6 is the *resonant frequency*. Label it with the symbol f_0. The resonant frequency is the frequency at which the impedance of the series combination of a resistor, capacitor, and inductor is minimum. This occurs at a frequency where the values of X_L and X_C are equal.

Prediction 2-7: On the axes above draw a curve that represents $X_L - X_C$ vs. frequency. Be sure to label it.

Prediction 2-8: Use your results from Predictions 2-5 and 2-7 to determine the general expression for the resonant frequency f_0 as a function of L and C. [**Hint:** You will need the expressions for X_C and X_L found in the box below.] Show your calculation below.

$$X_C = 1/(2\pi f C) \quad \text{and} \quad X_L = 2\pi f L$$

Test your predictions

Activity 2-1: The Resonant Frequency of a Series RLC Circuit.

1. Open the experiment file **RLC Resonance (L08A2-1).**

2. **Calibrate** the probes or **load the calibration,** if this has not already been done. **Zero** the probes with them disconnected from the circuit.

3. Connect the circuit with resistor, capacitor, inductor, signal generator, and probes shown in Figure 8-3.

4. Set the signal generator to a frequency of 10 Hz and amplitude of 5 V. (You should check this with the voltage probe.)

5. **Begin graphing.** When you have a good graph, **stop graphing** to capture the graph.

6. Use the **analysis feature** in the software to determine the peak voltage V_R^{max} and peak current I^{max}.

7. Enter the data in the first row Table 8-3.

Table 8-3

Frequency (Hz)	V^{max} (V)	I^{max} (A)
10		
15		
20		
25		
30		
35		
40		

8. Repeat steps 5–7 for the remaining frequencies in Table 8-3. Be sure that the amplitude of the signal generator is always 5 V. (You should check this with the voltage probe.)

Question 2-1: Does the behavior of the voltage across the resistor and current in the circuit in Figure 8-3 agree with your predictions, especially Predictions 2-5 to 2-7? Explain.

Prediction 2-9: Calculate the resonant frequency for your circuit. Show your calculations. [**Hint:** Use the formula from Prediction 2-8 and the actual values of the capacitance and inductance.]

$$f_0 = \underline{\qquad} \text{ Hz}$$

Prediction 2-10:   Calculate the values of the peak current through the circuit and the peak voltage across the resistor at resonance. Show your work and enter your results Table 8-4 in the column labeled Calculated. [**Hint:** Use the formula above for Z.]

Table 8-4

Resonant values	Calculated	Experimental
Peak current I^{max} (A)		
Peak voltage V_R^{max}(V)		

9. Measure the resonant frequency of the circuit to within 1 Hz. To do this, **begin** graphing and slowly adjust the frequency of the signal generator until the peak voltage across the resistor is maximum. (Use the results from Table 8-3 to help you locate the resonant frequency.)

$$f_0 = \underline{\quad\quad} \text{ Hz}$$

Question 2-2:   How does this experimental value for the resonant frequency compare with your Prediction 2-9?

11. Use the **analysis feature** to determine the peak voltage V_R^{max} and peak current I^{max} at the resonant frequency. Enter these values in Table 8-4 in the column labeled Experimental.

Question 2-3:   How do these experimental values for V_R^{max} and I^{max} compare with your Prediction 2-10?

Question 2-4:   List a few ways you might improve your calculation so the numbers you entered in the Calculated column might better match the actual results.

In a radio receiver, the signal generator is replaced by a long antenna, which picks up all of the radio signal frequencies. (This is similar to a microphone, mentioned in the previous investigation, detecting all the frequencies, or pitches, of music at once.) By strategically choosing values of C and L you can tune the circuit to the frequency of your favorite radio station, meaning the circuit is at resonance (the amplitude of the voltage across R is a maximum) for that particular station's frequency. (In many real radio receivers, a variable capacitor is used. When you turn the knob, you change the capacitance.)

Question 2-5: You tune your radio to a station broadcasting at 1010 kHz. Find the capacitance C necessary to make the resonant frequency of the circuit equal to the frequency of this radio station if $L = 800$ mH. Show your work.

If you have time, do the following extension.

Extension 2-2: Phase in an RLC Circuit

In Lab 7, you investigated the phase relationship between the current and voltage in an AC circuit composed of a signal generator connected to one of the following circuit elements: a resistor, capacitor, or an inductor. You found that the current and voltage are in phase when the element connected to the signal generator is a resistor, the current leads the voltage with a capacitor, and the current lags the voltage with an inductor.

You also discovered that the reactances of capacitors and inductors change in predictable ways as the frequency of the signal changes, while the resistance of a resistor is constant—independent of the signal frequency. When considering relatively high or low signal frequencies in a simple RLC circuit, the circuit element (either capacitor or inductor) with the highest reactance is said to "dominate" because this element determines whether the current lags or leads the voltage. At resonance, the reactances of capacitor and inductor cancel and do not contribute to the impedance of the circuit. The resistor then is said to dominate the circuit.

In this extension, you will qualitatively determine the phase relationship between the applied voltage (signal generator voltage) and current in an RLC circuit.

Consider the RLC circuit shown in Figure 8-4.

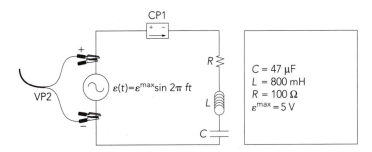

Figure 8-4: RLC series circuit.

Question E2-6: Which circuit element (the resistor, inductor, or capacitor) dominates the circuit in Figure 8-4 at frequencies well below the resonant frequency? Explain.

Question E2-7: Which circuit element (the resistor, inductor, or capacitor) dominates the circuit in Figure 8-4 at frequencies well above the resonant frequency? Explain.

Question E2-8: In the circuit in Figure 8-4, will the current through the resistor always be in phase with the voltage across it, regardless of the frequency? Explain your reasoning.

Prediction E2-11: In the circuit in Figure 8-4, will the current through the resistor always be in phase with the applied voltage from the signal generator? Why or why not?

Prediction E2-12: If your answer to Prediction E2-11 was *no*, then which will lead for frequencies below the resonant frequency (current or voltage)? Which will lead for frequencies above the resonant frequency (current or voltage)?

Test your predictions.

1. Open the experiment file called **RLC Phase (L08E2-2).**

2. **Calibrate** the probes or **load the calibration**, if this has not already been done. **Zero** the probes while disconnected from the circuit.

3. Connect the circuit shown in Figure 8-4.

4. Set the signal generator to a frequency 20 Hz below the resonant frequency you measured in Activity 2-1, and set the amplitude of the signal to 5 V.

5. **Begin graphing. Stop graphing** to capture the data when you have good graphs.

6. Determine whether the current or applied voltage leads.

Question E2-9: Which leads—applied voltage, current, or neither—when the AC signal frequency is lower than the resonant frequency? Were your predictions correct? Why or why not? Explain.

7. Set the signal generator to a frequency 20 Hz above the resonant frequency with the amplitude of the signal still 5 V.

8. Determine whether the current or applied voltage leads.

Question E2-10: Which leads—applied voltage, current or neither—when the AC signal frequency is higher than the resonant frequency? Were your predictions correct? Why or why not? Explain.

Prediction E2-13: Which will lead for an applied signal at the resonant frequency (current or voltage)?

9. Set the signal generator to the resonant frequency you measured in Activity 2-1, and set the amplitude of the signal to 5 V.

10. Determine whether the current or applied voltage leads.

Question E2-11: At resonance, does the current or applied voltage lead?

Question E2-12: Other than looking for a maximum peak current as the frequency is changed, can you suggest another method for experimentally locating the resonant frequency of an RLC circuit? Explain.

Summary: In Investigation 1, you learned how the frequency dependence of impedance for capacitors and inductors can be exploited to design filters to pass high- or low-frequency signals. Furthermore, you saw how this could be applied to the design of speakers. In this investigation you have investigated the phenomenon of resonance. One common application of the frequency-dependent impedances of inductors and capacitors is the series RLC resonant circuit. This circuit can be used, for example, as a radio tuning circuit.

HOMEWORK FOR LAB 8:
INTRODUCTION TO AC FILTERS AND RESONANCE

1. Carefully justify why a DC signal can be thought of as a 0-Hz AC signal.

2. A series RLC circuit has elements with the following values: $R = 10\ \Omega$, $L = 0.5$ H, and $C = 2\ \mu$F. What is the resonant frequency of this circuit? Show all work.

3. If you change the value of the resistor in Question 2 will the resonant frequency change. Explain.

4. Describe the magnitude of the impedance of a capacitor at very low and very high frequencies qualitatively.

5. The magnitude of the impedance of a capacitor, called the capacitative reactance, is represented by the symbol X_C. Based on your answer to Question 4, is the capacitative reactance related to the frequency f and capacitance C by the relation $X_C = 2\pi f C$ or $X_C = 1/2\pi f C$? (*circle one*) Explain your answer.

6. Describe the magnitude of the impedance of an inductor at very low and very high frequencies qualitatively.

7. The magnitude of the impedance of an inductor, called the inductive reactance, is represented by the symbol X_L. Based on your answer in Question 6, is the inductive reactance related to the frequency f and inductance L by the relation $X_L = 2\pi f L$ or $X_L = 1/2\pi f L$? (*circle one*) Explain your answer.

8. Describe precisely the behavior of an RLC circuit at the resonant frequency. Be quantitative as to (a) the size of the peak current as compared to the peak current at other frequencies, and (b) the phase difference between the applied signal voltage and the current.

APPENDIX: REALTIME PHYSICS ELECTRIC CIRCUITS EXPERIMENT CONFIGURATION FILES

Listed below are the settings in the *Experiment Configuration Files* used in these labs. These files are available from Vernier Software and Technology for *Logger Pro* software (Windows and Macintosh) and from PASCO for *Data Studio* (Windows and Macintosh). They are listed here so that the user can set up files for any compatible hardware and software package.

Experiment File	Description	Data Collection	Data Handling	Display
Current Model (L0A1-5)	Displays and graphs Current 1 and Current 2 vs. time.	25 points/sec Current probes 1 and 2 Digital and graphical display of inputs	NA	Two sets of graph axes with lines Current: −0.6 to +0.6 A Time: 0–10 sec
Two Voltages (L01A2-4a)	Displays and graphs Voltage 1 and Voltage 2 vs. time.	25 points/sec Voltage probes 1 and 2 Digital and graphical display of inputs	NA	Two sets of graph axes with lines. Voltage: −5 to +5 V Time: 0–10 sec
Current and Voltage (L01A2-4b)	Displays and graphs Voltage 1 and Current 2 vs. time.	25 points/sec Voltage probe 1 and Current probe 2 Digital and graphical display of inputs	NA	Two sets of graph axes with lines. Voltage: −5 to +5 V Current: −0.6 to +0.6 A Time: 0–10 sec
Two Currents (L02A1-2)	Displays and graphs Current 1 and Current 2 vs. time.	25 points/sec Current probes 1 and 2 Digital and graphical display of inputs	NA	Two sets of graph axes with lines. Current: −0.6 to +0.6 A Time: 0–10 sec
Batteries (L03A1-2)	Displays and graphs Voltage 1 and Voltage 2 vs. time.	25 points/sec Voltage probes 1 and 2 Digital and graphical display of inputs	NA	Two sets of graph axes with lines. Voltage: −5 to +5 V Time: 0–10 sec
Internal Resistance (L03A2-2)	Displays and graphs Voltage 1 and Current 2 vs. time.	25 points/sec Voltage probe 1 and Current probe 2 Digital and graphical display of inputs	NA	Two sets of graph axes with lines. Voltage: −5 to +5 V Current: −0.6 to +0.6 A Time: 0–10 sec
Ohm's Law (L03A3-1)	Displays and graphs Current 1 and Voltage 2 vs. time.	25 points/sec Current probe 1 and Voltage probe 2 Digital and graphical display of inputs	NA	Two sets of graph axes with lines. Voltage: 0 to +3 V Current: 0 to +0.5 A Time: 0–30 sec
Dependence of C (L05A1-2)	Displays data table and axes for Capacitance vs. Area (or Separation). Data can be entered into the table, and an active graph results.	Data are entered manually by double clicking on spaces in the data table	NA	One set of graph axes. Capacitance: 0–1.0 nF Area: 0–0.5 m^2 Seperation: 0–100 mm

Experiment File	Description	Data Collection	Data Handling	Display
Capacitor Decay (L05A3-1)	Displays and graphs Current 1 and Voltage 2 vs. time.	25 points/sec Current probe 1 and Voltage probe 2. Digital and graphical display of inputs	NA	Two sets of graph axes with lines. Voltage: −6 to +6 V Current: −0.6 to +0.6 A Time: 0–20 sec
LR DC Circuit (L06A1-1)	Displays Voltage 1 and Current 2.	Digital display of inputs	NA	Meter display of Voltage and Current
Switching Circuit (L06A1-2)	Displays and graphs Voltage 1 and Current 2 vs. time.	250 points/sec Voltage probe 1 and Current probe 2. Digital and graphical display of inputs	NA	Two sets of graph axes with lines. Voltage: −6 to +6 V Current: −0.1 to +0.1 A Time: 0–5 sec
Inductor Polarity (L06A2-1)	Displays and graphs Voltage 1 and Current 2 vs. time.	250 points/sec Voltage probe 1 and Current probe 2 Digital and graphical display of inputs	NA	Two sets of graph axes with lines. Voltage: −6 to +6 V Current: −0.1 to +0.1 A Time: 0–5 sec
RL Circuit (L06A3-1)	Displays and graphs Voltage 1 and Current 2 vs. time.	5000 points/sec Voltage probe 1 and Current probe 2 Digital and graphical display of inputs Triggered when Voltage probe 1 less than 4 V	NA	Two sets of graph axes with lines. Voltage: −6 to +6 V Current: −0.1 to +0.1 A Time: 0–0.2 sec
Time Varying Signal (L07A1-1)	Displays and graphs Current 1 and Voltage 2 vs. time.	25 points/sec Current probe 1 and Voltage probe 2 Digital and graphical display of inputs	NA	Two sets of graph axes with lines. Voltage: −6 to +6 V Current: −0.2 to +0.2 A Time: 0–5 sec
Signal Generator (L07A1-2)	Displays and graphs Voltage 1 vs. time.	500 points/sec Voltage probe 1 Digital and graphical display of inputs. Graphs in repeat mode	NA	One graph axes with lines. Voltage: −6 to +6 V Time: 0–0.6 sec
Resistor with AC (L07A2-1)	Displays and graphs Current 1 and Voltage 2 vs. time.	500 points/sec Current probe 1 and Voltage probe 2. Digital and graphical display of inputs. Graphs in repeat mode	NA	Two sets of graph axes with lines. Voltage: −6 to +6 V Current: −0.1 to +0.1 A Time: 0–0.5 sec
Capacitor with AC (L07A3-1)	Displays and graphs Current 1 and Voltage 2 vs. time.	500 points/sec Current probe 1 and Voltage probe 2. Digital and graphical display of inputs. Graphs in repeat mode.	NA	Two sets of graph axes with lines. Voltage: −6 to +6 V Current: −0.1 to +0.1 A Time: 0–0.5 sec

Experiment File	Description	Data Collection	Data Handling	Display
Inductor with AC (L07A3-3)	Displays and graphs Current 1 and Voltage 2 vs. time.	500 points/sec Current probe 1 and Voltage probe 2. Digital and graphical display of inputs. Graphs in repeat mode	NA	Two sets of graph axes with lines. Voltage: −6 to +6 V Current: −0.1 to +0.1 A Time: 0–0.5 sec
Capacitative Filter (L08A1-1)	Displays and graphs Current 1 and Voltage 2 vs. time.	500 points/sec Current probe 1 and Voltage probe 2. Digital and graphical display of inputs. Graphs in repeat mode	NA	Two sets of graph axes with lines. Voltage: −6 to +6 V Current: −0.1 to +0.1 A Time: 0–0.5 sec
Inductive Filter (L08A1-2)	Displays and graphs Current 1 and Voltage 2 vs. time.	500 points/sec Current probe 1 and Voltage probe 2. Digital and graphical display of inputs. Graphs in repeat mode	NA	Two sets of graph axes with lines. Voltage: −6 to +6 V Current: −0.1 to +0.1 A Time: 0–0.5 sec
RLC Resonance (L08A2-1)	Displays and graphs Current 1 and Voltage 2 vs. time	500 points/sec Current probe 1 and Voltage probe 2. Digital and graphical display of inputs. Graphs in repeat mode	NA	Two sets of graph axes with lines. Voltage: −6 to +6 V Current: −0.1 to +0.1 A Time: 0–0.5 sec
RLC Phase (L08E2-2)	Displays and graphs Current 1 and Voltage 2 vs. time	500 points/sec Current probe 1 and Voltage probe 2. Digital and graphical display of inputs. Graphs in repeat mode	NA	Two sets of graph axes with lines. Voltage: −6 to +6 V Current: −0.1 to +0.1 A Time: 0–0.5 sec